Approaches in Highly Parameterized Inversion: bgaPEST, a Bayesian Geostatistical Approach Implementation With PEST—Documentation and Instructions

By Michael N. Fienen, Marco D'Oria, John E. Doherty, and Randall J. Hunt

Groundwater Resources Program
Global Change Research and Development

Techniques and Methods, Book 7, Section C9

U.S. Department of the Interior
U.S. Geological Survey

U.S. Department of the Interior
KEN SALAZAR, Secretary

U.S. Geological Survey
Marcia K. McNutt, Director

U.S. Geological Survey, Reston, Virginia: 2013

For product and ordering information:
World Wide Web: http://www.usgs.gov/pubprod
Telephone: 1-888-ASK-USGS

For more information on the USGS—the Federal source for science about the Earth, its natural and living resources, natural hazards, and the environment:
World Wide Web: http://www.usgs.gov
Telephone: 1-888-ASK-USGS

Suggested citation:
Fienen, M.N., D'Oria, Marco, Doherty, J.E., and Hunt, R.J., 2013, Approaches in highly parameterized inversion: bgaPEST, a Bayesian geostatistical approach implementation with PEST—Documentation and instructions: U.S. Geological Survey Techniques and Methods, book 7, section C9, 86 p., available online at http://pubs.usgs.gov/tm/07/c09

Contents

Figures

Tables

Conversion Factors

Multiply	By	To obtain
foot (ft)	0.3048	meter (m)
gallon per minute (gal/min)	0.06309	liter per second (L/s)
cubic foot per second (ft^3/s)	0.02832	cubic meter per second (m^3/s)

Temperature in degrees Fahrenheit (°F) may be converted to degrees Celsius (°C) as follows:
$$°C = (°F - 32)/1.8$$

Approaches in Highly Parameterized Inversion: bgaPEST, a Bayesian Geostatistical Approach Implementation With PEST—Documentation and Instructions

By Michael N. Fienen[1], Marco D'Oria[2], John E. Doherty[3], and Randall J. Hunt[1]

Abstract

The application bgaPEST is a highly parameterized inversion software package implementing the Bayesian Geostatistical Approach in a framework compatible with the parameter estimation suite PEST. Highly parameterized inversion refers to cases in which parameters are distributed in space or time and are correlated with one another. The Bayesian aspect of bgaPEST is related to Bayesian probability theory in which prior information about parameters is formally revised on the basis of the calibration dataset used for the inversion. Conceptually, this approach formalizes the conditionality of estimated parameters on the specific data and model available. The geostatistical component of the method refers to the way in which prior information about the parameters is used. A geostatistical autocorrelation function is used to enforce structure on the parameters to avoid overfitting and unrealistic results. Bayesian Geostatistical Approach is designed to provide the smoothest solution that is consistent with the data. Optionally, users can specify a level of fit or estimate a balance between fit and model complexity informed by the data. Groundwater and surface-water applications are used as examples in this text, but the possible uses of bgaPEST extend to any distributed parameter applications.

Introduction

This report documents the theory and computer code of a Bayesian Geostatistical Approach (BGA[4]). "Bayesian" refers to theory used to solve the inverse problem; "Geostatistical" refers to the use of an autocorrelated spatial or temporal function to provide prior information about the parameters. This BGA approach has been coded by using the protocol and approach of PEST (Doherty, 2010a,b), the most widely used computer code of its type. Though BGA and PEST have been applied primarily to the environmental modeling field, a general view of the inverse problem discussed herein covers most classes of problems where measurements of a system are used to infer system properties that create the measured value. Thus, this bgaPEST formulation is meant to be applicable to any class of inverse problem that adheres to the concepts described herein. In this introductory section, we outline the conceptual framework of the method; theory, implementation, and instructions for use of the bgaPEST computer code version 1.0 follow this conceptual discussion.

Environmental modeling can facilitate informed management of natural resources. In most cases,

---

[1]U.S. Geological Survey, Wisconsin Water Science Center, Middleton, Wisconsin.

[2]Department of Civil and Environmental Engineering and Architecture, University of Parma, Parma, Italy.

[3]Watermark Numerical Computing, Brisbane, Australia and Australian National Centre for Groundwater Research and Training, Adelaide, Australia.

[4]Whereas the Bayesian Geostatistical Approach is referenced with the all capital abbreviation "BGA," lowercase is used in the software name ("bgaPEST") to visually differentiate "BGA" and "PEST."

models represent physical processes such as groundwater flow, contaminant transport, surface water flood routing, and so forth. Simulating the overarching physics and chemistry with governing equations and concepts such as conservation of mass and momentum is only part of what is required. Once a model adequately represents the processes of the problem, it is still only an abstract tool until estimates of the physical characteristics are tuned to observations from the specific area to be managed.

Input values that are used by processes are what control the behavior of a model at a given site; we refer to these input values as "parameters." They represent specific characteristics (for example, hydraulic conductivity, recharge rate, chemical decay rate) of the natural system at an area of interest. Representative values appropriate for parameters in natural systems are often difficult to measure directly, however. Observations of system state (for example, water levels, streamflow magnitudes, chemical concentrations) are often easier to measure, and the model is chosen so that the observations of the system correspond to output values from the model. The outputs of observation data are dependent to varying degrees on the values of parameters. Therefore, changes in the simulated system state that align with observations allow inference of parameter values not available by direct measurement. "Parameter estimation" and "calibration" are both terms describing the process of incorporating site-specific observations to inform parameters and salient processes of a model in order to improve the representativeness and predictive ability of the model.

A model, therefore, can be thought of as a data-processing tool that quantitatively tests conceptualizations of a system as well as a simulator of physical processes; metaphorically, data processing is a pipeline from field observation to model parameters and ultimately to more representative predictions for supporting management decisions.

Put in terms of a Bayesian description, this process is one in which the model is a vehicle for updating soft knowledge or expert knowledge of the system (called a priori understanding in the Bayesian context). This initial vehicle is filtered by a measure of its ability to simulate the natural world informed, or updated, by site-specific observations (called a

posteriori understanding). In the Bayesian approach, the ability to simulate features of the natural world "conditions" or narrows the wide range of possible outcomes that result from general expert-knowledge and soft knowledge alone. A key benefit of the Bayesian approach is that it provides a theoretically rigorous way to continually incorporate new information and, in turn, update a posteriori understanding.

Purpose and Scope

This report is intended to serve two purposes. First, a Bayesian approach to parameter estimation—expressed in the context of the Bayesian Geostatistical Approach (BGA)—is described to provide an accessible and general tool for moving a model from a general simulator of a physical process to a more optimal tool, one that is tuned to a set of calibration information, which, in turn, can be used for improved prediction and decision-making. Second, a computer code—bgaPEST version 1.0—is introduced and documented in which BGA is deployed by means of the protocols and input/output concepts developed in the free and open-source PEST suite of software Doherty (2010a). This report gives details on the mathematical theory behind BGA, followed by detailed instructions for using the computer program. Conventions and assumptions for using the program also are included in the discussion. To our knowledge, this application marks the first implementation of a general BGA code available for widespread use.

The bgaPEST input framework is consistent with the input block and keyword concepts described by the JUPITER project Banta and others (2006). Although the relation of design concepts is beyond the scope of this report, the input block and keywords needed to run bgaPEST are described fully in appendix 1. A full description of the format of the general approach of template and instruction files is omitted here; detailed descriptions are provided in the PEST documentation (Doherty, 2010a, chapter 3). All options implemented in template and instruction files in PEST are available in bgaPEST. A distributed parameterization scheme discussed in this report facilitates the introduction of flexibility to the model. This parameterization scheme also can be applied to any region of interest and at the extreme, where sufficient data are available, can allow a modeler to estimate a unique parameter values for each model node or cell. This level of detail leads to a large number of parameters, a condition that poses computational challenges; alleviating these computational challenges is an active area of ongoing research and thus is not covered in detail here.

This report includes an overview of theory and use of the bgaPEST code in the main text. Detailed input instructions for bgaPEST version 1.0 are in appendix 1, and quick-start instructions are in appendix 2. A detailed mathematical derivation of the BGA methods is in appendix 3, and example problems are in appendixes 5 through 7.

Obtaining the Software

The software for bgaPEST Version 1.0 is available for download at http://pubs.usgs.gov/tm/07/c09. This location includes a copy of this document and both executables and source code. As development of the code continues, a repository at http://github.com/mnfienen-usgs/bgaPEST provides a link to revisions in progress and provides a collaborative open-source space where users may submit revisions for consideration by the authors. As further development takes place, new code releases will be posted at http://pubs.usgs.gov/tm/07/c09.

The Bayesian Geostatistical Approach

The presentation of BGA is in two parts, first primarily as narrative then later as a detailed mathematical approach in appendix 3. Those most interested in simply applying bgaPEST to their specific problem will likely spend most time with the narrative. In both presentations, the concept of conditionality is fundamental. This concepts is expressed here as Bayes' theorem,which forms the foundation of the techniques described in the rest of this report:

$$p(\mathbf{s}|\mathbf{y}) \propto L(\mathbf{y}|\mathbf{s})\, p(\mathbf{s}) \qquad (1)$$

where: $\mathbf{s}$ is an $m \times 1$ vector of m parameter values, $\mathbf{y}$ is $n \times 1$ vector of n observations, $p(\cdot)$ indicates a probability density function (pdf), $L(\cdot)$ indicates a likelihood function, and $|$ indicates conditionality. Put into words, Bayes' theorem states that the posterior probability of parameters conditional on the observations $p(\mathbf{s}|\mathbf{y})$ (often referred to simply as "the posterior probability of $\mathbf{s}$ given $\mathbf{y}$") is proportional to the prior probability of the parameters $p(\mathbf{s})$ updated with the likelihood function $L(\mathbf{y}|\mathbf{s})$ that expresses how well $\mathbf{y}$ is estimated by using the model and a candidate parameter set $\mathbf{s}$. The pdfs in all cases are assumed to follow or at least be well-approximated by Gaussian distributions. This assumption is important and somewhat restrictive, but is made for computational simplicity. Active research is ongoing on alternatives to this approach, but the traditional Gaussian assumption is adopted in this report and is still considered a practical and useful assumption for many cases.

In the Bayesian context, expressing the parameters and the likelihood function as probability distributions formally incorporates an estimate of their uncertainty rather than treating the parameters as perfectly known values. All a posteriori (also called posterior) distributions are conditional upon the specific data used in the calibration process. Perhaps less obviously, posterior distributions are also conditional on all other modeling and data assumptions and decisions that go into formulating the problem: which model and what model options are chosen, numerical considerations such as discretization and solver convergence criteria, boundary conditions that may or may not be considered static and known, variance values and weights given to individual observations and parameters, and others. As a result, if any of these underlying assumptions and decisions change, it is expected in the Bayesian context that the parameters estimated and the associated posterior uncertainty also will change.

The conditionality includes *all* decisions made in the process of constructing a model and incorporating data;soft knowledge however, not all of this information is explicitly addressed by the modeler. In fact, the only *explicit* conditionality is on the observation data. The information contained in the prior pdf ($p(\mathbf{s})$) in equation 1 is critical because it represents the state of knowledge about the parameters in the system prior to updating through the calibration process. In this report, assumptions made prior to calibration are intentionally limited and are restricted to the assignment of a mean value (unknown) of each distinct parameter type and region within the model domain (beta association, described below) and a characteristic about continuity or smoothness of the parameter field (implemented as a variogram, also described below). The degree to which this continuity characteristic is enforced is dictated by the observations included and the subsequent performance of the model. This construction is similar to (and in some cases, mathematically equivalent to) Tikhonov regularization (Tikhonov, 1963a,b; Aster and others, 2005). Limitation of the information assumed a priori is similar to assuming a low or diffuse level of a priori soft knowledge (called an ignorance prior in Jaynes and Bretthorst, 2003, chapter 12). In this case, the resulting model is driven more by information obtained from site-specific observations than from prior assumptions based on soft knowledge. A goal of this approach is to limit the subjective information and to favor instead an objective and repeatable result based on observation data. Additionally, so-called structural parameters that enforce the characteristic smoothness are estimated. An algorithm that encompasses all of these aspects is considered an Empirical Bayes approach.

A Note on Parameters

The term "parameters" refers to discrete values of system state. When we describe "parameter type" we mean a group of parameters that belong to the same class of system state (for example, hydraulic conductivity or recharge). In applications appropriate for bgaPEST, there must be multiple parameter values of a given type that are spatially or temporally distributed. When a "parameter" is listed, the meaning is restricted to a single value of system model input that is to be estimated.

The concept of a beta association is important and is a concept that is revisited throughout this report. Fienen and others (2009) describe the need to to represent the generalized mean value for a specific parameter type (for example, hydraulic conductivity in a groundwater model) in a specific region of a model referred to as a "facies association." To be more general, this concept is incorporated here by the use of the term "beta," derived from the mathematical symbol used. In the methodology described in this report, parameter values are estimated by estimating a mean value (termed β in the mathematics) and the fluctuations about that mean. Each parameter, therefore, must be associated with a mean value.

It would be tempting to use another term such as "zone" or "facies" to describe this concept, but the term "beta association" was selected specifically to highlight the flexibility of the concept. The important idea is that the method described in bgaPEST depends on being able to associate each parameter with a mean value. In the case of distributed parameters (for example, hydraulic conductivity or recharge being distributed throughout a region in a model in which each model cell or node is assigned a unique value), the subdivision of the entire model domain into beta associations accounts for hydrogeologic contacts or facies to be delineated. This delineation assumes that there is little or no correlation across these natural divisions. Similarly, parameters of one type are typically not correlated with parameters of a different type. Beta associations allow the inclusion of multiple parameter types and the delineation of important geologic features in distributed parameter sets. In the BGA algorithm, parameters in different beta associations are assigned zero correlation.

The likelihood function in equation 1, $L(\mathbf{y}|\mathbf{s})$, expresses the correspondence of model outcomes with field observations colocated in space and time. This correspondence is expressed as the sum of the squared differences between outcomes and observations, weighted by a covariance matrix, which expresses the relative certainty of each observation. This is equivalent to the weighted measurement objective function in PEST (Doherty, 2010a). The advantages to the Bayesian approach stem from the conceptual framework, the ability to use a probability density function to represent parameterization rather than single values, and the empirical nature of the balance between prior information and likelihood. These elements of Bayes' equation form the fundamental basis for the bgaPEST software described here.

The *geostatistical* aspects of the method are

expressed in the prior pdf ($p(\mathbf{s})$) of the parameter values. Geostatistics is a form a interpolation that uses a spatial function called a variogram to fill in information between data points. The most common technique of geostatistics is called kriging. In this section, we discuss geostatistics in a conceptual way using a photographic image as an example. The mathematics relevant to BGA are discussed in later sections. More details about the history and use of geostatistics in general, including software for using the technique are found in Isaaks and Srivastava (1989), Deutsch and Journel (1992), Kitanidis (1997), and Remy and others (2009).

The photograph in the left panel of figure 1 is a JPG image taken on a digital camera. As a grayscale image, the information of the image can be stored in a matrix with the number of rows and columns, 320 rows and 240 columns in this case, indicating the number of pixels in the vertical and horizontal dimensions, respectively. The values at each pixel are a brightness value, in this case normalized to a maximum value of 64.0. In the original image, there are $320 \times 240 = 76,800$ pixels, each of which may be considered a discrete packet of information. To illustrate the kriging process, the photograph was first subsampled on an evenly spaced grid of 30 rows and 20 columns with the brightness value retained at each location. This subsampling results in a greatly reduced set of information containing $30 \times 20 = 600$ pixels. In the photograph in the right panel of figure 1, the faint impression of the subsampling grid is visible in some areas as the subsampled values are depicted at those locations. Using the geostatistical technique of kriging with an appropriate variogram, it is possible to "fill in" the missing data between subsampled data points to present a full image of the matrix but with substantially less detail than the original.

The main role of the variogram in geostatistics and, indeed, in BGA, is to act as a constraint, controlling the shape of the interpolated values filling in between the known data values. In BGA, this connection is not quite as direct as in the photograph interpolation example, but it is useful to think of the variogram (as a quantification of the prior pdf) as a control on the shape of the estimated parameters.

Various other interpolation techniques could be used to fill in the missing data and each would have its own degree of information loss or smoothing relative to the original. Kriging has a long history of use in earth science applications and, although the interpolated photographic image in the example is much smoother than the original image, there is more shape and information than if, for example, linear interpolation had been used to fill the values between each data point.

The variogram used in kriging is an empirical function that characterizes the difference between a property as a function of separation distance. To determine a variogram appropriate for a problem, the first step is to plot a variogram function (a function of difference in property value) against separation distance (depicted as red "x" marks in figure 2). Next, a function type is selected from a family of valid variogram model types. In the photograph example an exponential variogram is used. Later in this report, more mathematical details about variograms and variogram choice are presented.

Returning to the geostatistical aspect of BGA, a variogram model is used as the prior information in the Bayesian construction. As discussed above, a goal of BGA is to specify little information in the prior and to allow the information contained in the calibration dataset to inform the results as much as possible. This is accomplished by specifying only the family of variogram model used rather than specifying its specific shape. Using the example in figure 2, the family of variogram (in this case exponential) indicates only that the function will assume a curvilinear shape; the specific parameters or the variogram function dictate the rate of curvature. In terms of using a variogram for the prior distribution in BGA, specifying the variogram type informs only the most general characteristic of the field (for example, the field must be continuous and "smooth"). The degree to which this characteristic is enforced is controlled by the calibration dataset.

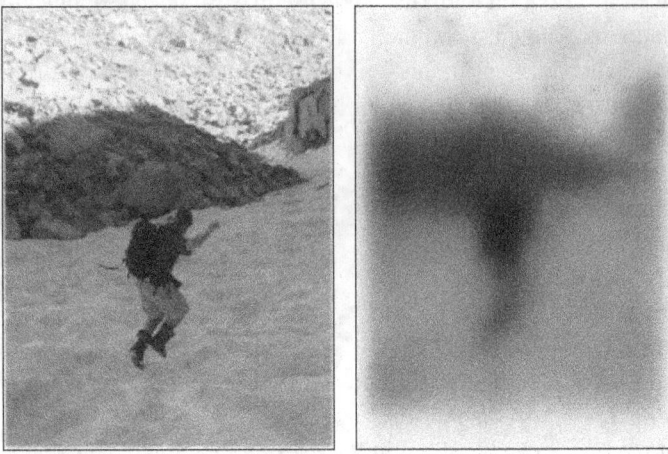

Figure 1.   Photographic images illustrating the geostatistical interpolation process. The image on the left is a JPG format image with 320×240 pixels. The image on the right is also a JPG image, but it was created by first sampling a subset of the pixels in the original image (30 in the vertical direction and 20 in the horizontal) and then using the geostatistical technique of kriging to interpolate values for the other pixels. The interpolation was done by using the sGeMS software package (Remy and others, 2009).

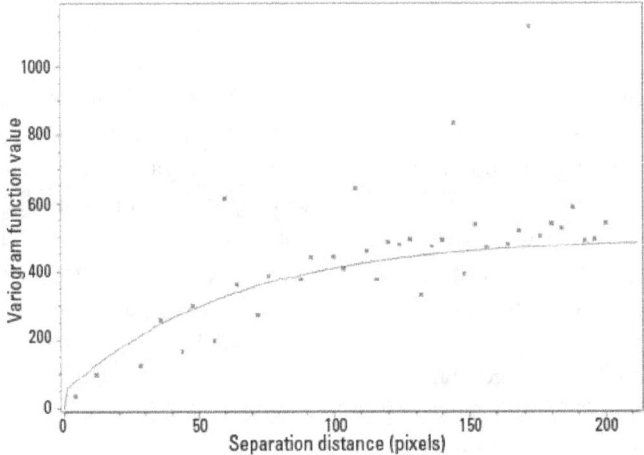

Figure 2.   Exponential variogram fit to empirical variogram for the image processing example. The empirical variogram binned values are depicted by red "x" marks whereas the continuous black line shows the analytical variogram fit to the empirical values. The fit was done manually by using the sGeMS software package (Remy and others (2009)).

Overview

The Bayesian geostatistical approach is described in detail by Kitanidis and Vomvoris (1983), Hoeksema and Kitanidis (1984), Kitanidis (1995), and Nowak and Cirpka (2004) among others. This section is a conceptual overview of the method. A more detailed description, including mathematical details, is in appendix 3.

The core of the Bayesian geostatistical inverse method is Bayes' theorem, which states

$$p(\mathbf{s}|\mathbf{y}) \propto L(\mathbf{y}|\mathbf{s})p(\mathbf{s}) \tag{2}$$

where $\mathbf{y}$ are the measured data, $\mathbf{s}$ are the unknown parameters, $p(\mathbf{s}|\mathbf{y})$ is the posterior probability density function (pdf) of $\mathbf{s}$ given $\mathbf{y}$, $L(\mathbf{y}|\mathbf{s})$ is the likelihood function, and $p(\mathbf{s})$ is the prior pdf of $\mathbf{s}$. Details of these pdfs are explained below.

Figure 3 depicts one-dimensional distributions graphically illustrating equation 2. In this example, the prior distribution $p(\mathbf{s})$ is diffuse, meaning the variance is relatively high and, correspondingly, commitment to a particular value is low. The likelihood function $L(\mathbf{y}|\mathbf{s})$, on the other hand, has lower variance, suggesting a process that brings a higher level of certainty to the estimation of the parameters $(\mathbf{s})$ than is indicated by the prior distribution only. The resulting posterior distribution $p(\mathbf{s}|\mathbf{y})$ is a convolution of the prior and likelihood functions. The peak is shifted significantly from the prior toward the likelihood and is narrower, representing less uncertainty.

In bgaPEST, an empirical Bayes perspective (Robbins, 1956; Casella, 1985) is adopted. Empirical Bayes means that the general characteristics of the prior and (optionally) epistemic covariances introduced above are provided in the model setup, but the values of "structural" parameters that control the structure of the system—the balance between smoothness and misfit—are estimated from the observation data. In other words, the level of roughness in the solution is dictated by the information content of the observation data rather than specified by the user ahead of time.

The prior distribution is the main mechanism by which soft knowledge about the parameter field is imparted on the parameter estimation process. In the Empirical Bayes perspective, this soft knowledge is intentionally limited such that significant flexibility is available to the algorithm and a specific practitioner's preconceived notions, which are more subjective, are replaced by the objective power of the site-specific observations. This idea is also inspired by Chamberlin's concept of multiple working hypotheses (Chamberlin, 1890). Chamberlin warned of scientists falling victim to a "paternalistic affection" for their initial explanation of a phenomenon such that they are blind to other explanations that may be more appropriate. This is not to discount the value of soft knowledge—indeed, the general characteristics imparted through specification of the prior information and the interpretation of the results of using BGA rely deeply on expertise—but it highlights a goal of leaving as much flexibility as possible in the process.

In bgaPEST, then, the practitioner specifies a type of variogram (nugget, linear, or exponential) that is used to control the variability—smoothness or roughness—of parameters within a beta association, but the *degree* to which this characteristic is enforced is determined by a Bayesian adaptation of restricted maximum likelihood (RML). In RML, the value of structural parameters that control the variogram behavior is treated as a probability distribution and the most likely values resulting in either the best possible fit (if the epistemic error term is estimated) or a user-specific level of fit (if the epistemic error term is fixed) are estimated. The Bayesian adaptation to RML in this report is through the inclusion of prior information and uncertainty, which is not strictly possible in traditional RML. "Fit," in this context, refers to the correspondence between observation data and model outputs colocated in space and time with the measured observations. Fit and epistemic error are discussed in more detail in the next section. A danger of providing a model with substantial flexibility is an "overly complex" model that is "overfit" (for example, Draper and Smith, 1966; Hill, 2006). To mitigate this issue, the RML approach is consistent with the principle of maximum entropy such that the smoothest solution is chosen, an approach based on the structural parameters estimated from the data. For a discussion of subtle formal differences from minimum relative entropy, see Rubin (2003, p. 333–342).

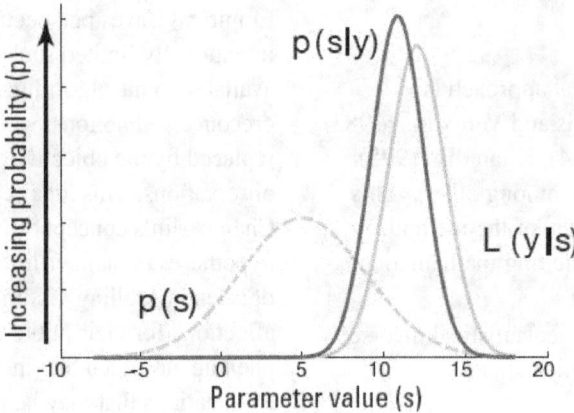

Figure 3.   Graphical illustration of Bayes' theorem.

An extension of this approach is the inclusion of information about the prior mean (Nowak and Cirpka, 2004). Although the mean is estimated in the solution, a prior value and covariance can be supplied to constrain the estimate. Typically, a relatively high covariance magnitude is used so that the constraint on the estimated mean is weak or "diffuse." Thus the prior mean principally serves the role of providing numerical stability rather than compelling the solution to adhere closely to prior values. Similarly, prior information and covariance can be supplied on the structural parameters to constrain the estimated values to more closely follow an initial conception of the parameter field variability.

The forward model is constructed to generate outputs of values colocated in space and time with measured observations. The likelihood function quantifies the difference (misfit) between the model simulated outputs and associated observations. In all modeling, perfect correspondence between forecasts and observations is neither attainable nor desirable. The observations themselves are corrupted by measurement errors, and perfect correspondence between the exact nature of the measurements and the simulated counterparts usually is lacking. This corruption is due to uncertainty from sources including the paucity of observations, imperfections in the conceptual model, and approximations made to codify the physics of the phenomena into a numerical model framework. All of these sources of uncertainty are described by the overarching term "epistemic uncertainty" (Rubin, 2003, p. 4). This epistemic uncertainty characterizes the expected misfit between

simulated and observed equivalents, and is expressed through a covariance function. As a result, the likelihood function can be characterized by a Gaussian distribution with zero mean and covariance defined by the epistemic uncertainty.

Structural Parameters

The term "structural parameters" used here has a specific meaning. Similar to the more general term "parameters," structural parameters are variable values that are estimated in the bgaPEST algorithm. Unlike typical parameters, however, structural parameters do not *directly* control physical aspects of the system in the way that, for example, hydraulic conductivity or stream roughness do in hydrologic models. Instead, structural parameters control the structure of the general parameters. For example, the variogram values (for example variance, slope, and correlation length) that control the roughness of distributed parameter fields are structural parameters, as is the value of variance controlling epistemic uncertainty. Because these parameters must be estimated but are not directly connected to the physics of the problem, they are also referred to by other authors as "nuisance" parameters or "hyperparameters." We adopt the term "structural" to highlight the fact that the impact these parameters has on the solution is control of the shape or structure of the distributed parameter fields.

With both the prior pdf and likelihood function expressed as Gaussian distributions, the resulting posterior pdf also is Gaussian. The values of the parameters **s** that result in the maximum value of the posterior pdf are therefore the most likely solution on a point-by-point basis. The solution as a whole is always a somewhat smoothed version of reality, but the influence of small-scale variability can be approximated through conditional realizations. The balance between the strength of smoothing and the level of fit between simulated and observed equivalents is found through calculation of optimal values for the structural parameters. Optionally, this can include a value to quantify the epistemic uncertainty. The result will favor smoothness, but it may achieve a level of fit corresponding to an unrealistically low level of epistemic uncertainty. Hence, it is generally most appropriate to fix the level of epistemic uncertainty but allow the other structural parameters to be estimated.

Beta Associations

In an idealized problem, a single covariance model (for example, a single variogram) is flexible enough to encompass the entire variability of the hydraulic parameters. In many hydrologic applications, however, lithologic contacts and unconformities can create discontinuities in parameter values that a single covariance model cannot characterize. Partitioning the field either on the basis of data (for example, Fienen and others, 2004) or through interrogation of preliminary solutions (for example, Fienen and others, 2008) can greatly improve the parameter estimation results. This partitioning is implemented by imposing discontinuities in the stochastic field that censor correlation among all cells that do not occur in the same partition. In this context, "stochastic" refers to the entity being partitioned (namely, the correlation structure of the parameter field) but we emphasize here that the locations of the imposed discontinuities are themselves considered deterministic and certain. This concept of partitioning is consistent with zonal boundaries in models made up of homogeneous zones but it allows more flexibility by allowing properties within the zone to vary. Furthermore, multiple types of parameters (for example, hydraulic conductivity and porosity in a flow and transport model) are commonly estimated. Although these parameters may be related at the physical level, they

must correspond to different mean values, so similar censoring of correlation among different types of parameters also is necessary in most applications through partitioning.

For hydrogeologic applications, the term "facies association," from the facies architecture field, is an apt description for these partitions (Fienen and others, 2009). The term "facies association" typically refers to descriptive properties of a subset of a medium in the field or at least for a specific project. "Architectural elements" is used in the broader case where the characteristics are more formally defined (see Collinson (1969); Walker (1984, 1992); Swift and others (2003)). It would be appropriate to use the less restrictive and less transferable term "facies association" in hydrogeologic applications because when we subdivide the correlation structure of the medium, we often base the stochastic discontinuities (bounding surfaces, or contacts) on perceived hydraulic properties. These properties will often coincide with differences in age, provenance, or depositional environment, but such coincidence is not required for or by their use. In all cases, partitioning into facies associations is most effective when based on readily observable hydrologic or lithologic attributes.

For bgaPEST to be a more general tool (not limited to hydrogeologic modeling), we have broadened this concept by adopting the term "beta association." As shown in equation 3.2, the Greek letter β stands for the mean of a region of distributed parameters. Beta associations can, therefore, delineate regions of a distributed parameter field that have similar statistical properties and correspond to the same mean value; however, importantly, beta associations also can refer to completely different parameter types (for example, hydraulic conductivity and recharge).

To clarify our terminology, partitions delineated by stochastic discontinuity within a distributed parameter field are referred to as "beta associations," whereas zones of piecewise continuity are referred to herein as "homogeneous zones." The beta associations delineate sub-regions of the model domain that share correlation characteristics and are uncorrelated from neighboring beta associations; they are usually delineated by features that are easily identified in measured data or geologic conceptualizations of a given site area. In beta associations, variability of parameter values within each cell is allowed and constrained by the a priori covariance structure, whereas in homogeneous zones, a single parameter value represents the property for the entire zone. Beta associations also delineate regions in the model (whether defined by one or more parameter values) that correspond to different mean values (β).

Beta Associations and Zones: *Why aren't "beta associations" just called "zones"?*
Beta associations are a term specific to bgaPEST. As discussed in the main text, this term evolved from the term "facies association," which describes partitioning of parameter fields on the basis of hydrogeologic characteristics. This term was used in place of "zones" because of a long history of zones referring to regions of piecewise constant homogeneity (one parameter value applied to every node within a region). Beta associations are *not* homogeneous, so a distinct term was sought that describes the characteristic of regions partitioned according to their characteristics and the way these characteristics correlate to regions around them. To generalize beyond hydrogeologic applications, and to account for the fact the distinct parameter types require distinct partitions, "beta associations" was the term chosen. Each of these parameter partitions has a distinct mean value (β) to be estimated within the region, so the partitioning of the problem results in different β values; because the parameter type and/or region must be associated with a mean value (β), we use the term "beta associations."

Overview of bgaPEST

The use of BGA concepts described previously has been restricted to primarily academic and/or custom applications, owing to the case-by-case nature of the BGA coding. The BGA formulation used in bgaPEST is meant to make the approach generally available to a wider class of modeling problems. This generality is achieved by way of the following design considerations. The input/output design of bgaPEST follows that of the widely used PEST software (Doherty, 2010a,b). This approach has two primary restrictions. First, input provided to the model, and output derived from a model, uses an ASCII text file format. This restriction can be relaxed, however, provided that a translation utility can be deployed for converting data of another format—for example, binary—to or from ASCII, as appropriate. Second, the model must run in "batch" mode where many model runs can be called by PEST without user intervention. Therefore, the only kind of model that PEST cannot easily accommodate is one in which any changes to model input or the reading of model output must take place in a graphical user interface. This generality of model compatibility is a powerful capability that bgaPEST is able to exploit by virtue of efficient open-source modules that make this external control of a model possible using the same protocols as PEST.

As discussed below, bgaPEST must control the model for two purposes: to evaluate the likelihood function (assessing the correspondence between model output and colocated observation data, given a candidate set of parameter values) and to calculate the "Jacobian" or "sensitivity" matrix that is required for solving the calibration equations. To enable PEST (and bgaPEST) to write input for a model, template files are created that map named parameters into their proper place in input files for the model. More than one template file can be used corresponding to multiple model input files. To enable reading of output files, instruction files are created that contain a set of instructions (including locating specific line numbers or searching for specific text) that enable extraction of output values to be compared with site observation data. Leveraging the modules that implement the PEST input/output protocols takes advantage of the flexibility and generality of PEST. It also makes it possible to take advantage of certain utility programs already created to be compatible with the PEST suite of software. Programs created using the JUPITER program employ a very similar set of protocols by virtue of the PEST modules having been provided to the JUPITER project. As a result, template and instruction files created to work with a model are largely interchangeable among projects implemented in PEST, bgaPEST, and programs created using JUPITER. A full description of the format of template and instruction files is not within the scope of this report: detailed descriptions are provided in the PEST documentation (Doherty, 2010a, chapter 3). All options implemented in template and instruction files in PEST are available in bgaPEST.

This initial implementation of bgaPEST is written in Fortran 90. The calculation of the Jacobian (sensitivity or derivatives) matrix can be implemented either using a script written by the user or employing a Python[5] script provided with bgaPEST. The Python script depends on several utilities that are standard with PEST and available for download at http://www.pesthomepage.org. The necessary executable files are also provided with bgaPEST. For users on the Windows[6] operating system, installation of Python is optional because the Python codes are compiled into executables using py2exe that can be called by the main program. For users on Macintosh[7] or Linux[8] systems, all the code must be compiled for the native platform and Python should already be installed so the Python scripts may be called directly without need to compile them separately. The use of external derivatives (sensitivity) calculation with PEST and Python can be replaced by using the parallel external derivatives capabilities described in appendix 4. Alternatively, it would be possible to implement the general parallel run management suite (GENIE, Muffels and others (2012)).

Running bgaPEST

The bgaPEST program uses a single input control file in combination with template and instruction files to control the underlying model, and it generates several output files. These files are discussed in the context of progression of the bgaPEST program in the remainder of this section. Figure 4 shows the general progression of a bgaPEST parameter estimation run. The entire process is controlled by variables in the input .bgp file discussed below.

To obtain an optimal solution of the parameter estimation problem, multiple iterations are necessary. An iteration is defined as a single run of the entire estimation process with a particular set of values. Multiple iterations are required because of the nonlinearity of the problem and the necessity of estimating structural parameters separately from model parameters. Appendix 3 gives more detail about the methods used to obtain a solution for a set of optimal parameters and structural parameters in bgaPEST.

Outer iterations (also called BGA iterations) are wrapped around the traditional parameter estimation process with values of the structural parameters held constant. Inner parameter estimation iterations are performed to account for the (restricted maximum likelihood) estimation of structural parameters. If structural parameters are not chosen to be estimated, then a single outer iteration is performed using the initial values of structural parameters and inner iterations are performed until convergence or until the number of iterations reaches it_max_phi. If a line search (discussed below) is requested, this is performed within the inner iterations. If structural parameter optimization is requested, it is performed after convergence has occurred or maximum inner iterations have been reached. Then, restricted maximum likelihood is performed to estimate a new set of structural parameters. The interdependence between structural parameters and model parameters requires reiteration of the inner iterations and structural parameter estimation until either both have converged or the maximum number of outer iterations has been reached. At the end of both inner and outer iteration convergence, or exceedance of maximum iterations, posterior covariance is calculated, if requested.

Control Variables

Two types of variables are used in bgaPEST: control variables and data variables. Whereas data variables are values such as model parameters, observations, file names, and other data that are needed by the bgaPEST program, control variables, drive the actions that are performed on these data elements. As such, control variables operate on a different level from data variables and control either

---

[5]"Python®" is a registered trademark of the Python Software Foundation.

[6]"Windows®" is a registered trademark of the Microsoft group of companies.

[7]"Macintosh®" is a registered trademark of Apple, Inc. in the U.S. and other countries.

[8]"Linux®" is the registered trademark of Linus Torvalds in the U.S. and other countries.

Outer Iteration (maximum iterations defined by `it_max_bga`)
The outer iterations iterate until convergence of both
PHI and Structural Parameters (optional)

    Inner Iteration (maximum iterations defined by `it_max_phi`)
      The inner iteration progresses until convergence of PHI conditional
      on the current values of Structural Parameters
    end Inner Iteration

    If `linesearch == 1`
      Linesearch (maximum iterations defined by `it_max_linesearch`)
        The linesearch is an optional correction that can get optimization
        for PHI back on track if it strays
      end linesearch

    If `struct_par_opt ==1`
      Structural Parameter Optimization
      (maximum iterations defined by `it_max_structural`)
        If, optionally, Structural Parameter Optimization is invoked,
        iterations progress until convergence conditional on
        current process parameter values
      end Structural Parameter Optimization
end Outer Iteration

If `posterior_cov_flag == 1`
    Calculate posterior covariance
      posterior covariance is only calculated for the final parameter values

Figure 4.   Abbreviated flowchart showing the progression of the major bgaPEST procedures. Text in blue italics is interpretive, summarizing the more programmatic language represented in black plain type.

the reading/writing of data or the progression of the algorithm. Many control variables are straightforward (for example, `it_max_phi`, an integer, is the total number of iterations allowed in each quasi-linear inner estimation optimization; *default=10*). Such variables are defined in the context of the input instructions listed in appendix 1. Other control variables, however, are accompanied by important conventions regarding their impact on the performance of the algorithm. More detail is given about certain control variables in this section for these cases.

`structural_conv` *float, default=0.001*
    Convergence criterion for structural parameter convergence. Positive or negative values can be used to trigger two different measures of convergence, as noted below. In either case, however, convergence is compared to the absolute value of `structural_conv`.
    If positive, convergence is based on the absolute difference in structural parameter

objective function over consecutive iterations.

$$\mathrm{conv} = \mathrm{abs}\left(\Phi_{S,i} - \Phi_{S,i-1}\right) \qquad (3)$$

where i is the current structural parameter optimization iteration, $i-1$ is the previous structural parameter optimization iteration, and Φ_S is the structural parameter objective function.

If negative, convergence is based on the norm of the difference between consecutive structural parameter values.

$$\mathrm{conv} = \sqrt{\left(\frac{\theta_{i-1} - \theta_i}{\theta_{i-1}}\right)^T \left(\frac{\theta_{i-1} - \theta_i}{\theta_{i-1}}\right)} \qquad (4)$$

where i and $i-1$ are as defined above, and θ is a vector containing all structural parameters currently being estimated (may include epistemic uncertainty, if requested).

`phi_conv` *float, default=0.001* Convergence criterion for objective function convergence. The convergence at each inner iteration is evaluated as the absolute difference from one inner iteration to the next. This is evaluated as

$$\text{conv} = \text{abs}\left(\Phi_{T,i_{in}} - \Phi_{T,i_{in}-1}\right) \quad (5)$$

where i_{in} is the current inner iteration and Φ_T is the total objective function (equation 3.30).

bga_conv *float, default*=10×phi_conv
Convergence criterion for objective function outer iterations. The convergence at each outer iteration is evaluated as the absolute difference from one outer iteration to the next. This accounts for convergence of Φ_T and Φ_S. The convergence is evaluated as

$$\text{conv} = \text{abs}\left(\Phi_{T,i_{out}} - \Phi_{T,i_{out}-1}\right) \quad (6)$$

where i_{out} is the outer iteration and Φ_T is the total objective function (equation 3.30).

Q_compression_flag *integer, default*=0 Flag to determine how to calculate Q_{ss}. [0] = no compression—calculate full Q_{ss} matrix, [1] = calculate separate Q_{ss} matrix for each beta association. In addition to controlling the behavior of prior covariance compression, this flag also determines whether a full posterior covariance matrix or only the diagonal is calculated.

posterior_cov_flag *integer* Flag to determine whether posterior covariance matrix should be calculated. [0] = do not calculate posterior covariance matrix, [1] = calculate posterior covariance. If Q_compression_flag=1, only the diagonal of the posterior covariance matrix is calculated. If posterior_cov_flag=0 then 95 percent confidence intervals are not calculated and the output file <casename>.bpp.fin discussed below does not include confidence intervals.

Input Files

The bgaPEST program is run from the command line by typing bgaPEST.exe <casename>.bgp, where <casename> is a filename containing input instructions. Detailed input instruction are in appendix 1.

Output Files

Several output files are generated throughout the progression of a single bgaPEST run. These files are summarized in this section.

Record File

The main output file for bgaPEST is called <casename>.bpr. Initial values of bgaPEST input are repeated to form a record for the bgaPEST run. After each inner iteration, as defined above, the objective function is reported and external files are written that include current parameter values and observation values. After each outer iteration, structural parameter values also are reported for each beta association in which structural parameter estimation was requested and for the epistemic uncertainty term, if requested. After the final outer iteration, all structural parameter values—including those which were not estimated—are reported to make a complete record.

Parameter Value Files

The parameter values are written to files called <casename>.bpp.<#Oi>_<#Ii> where <#Oi> is the outer iteration number and <#Ii> is the inner iteration number. These ASCII files are printed in columns with the following headers: ParamName; ParamGroup; BetaAssoc; ParamVal. At the beginning of a bgaPEST run, a file <casename>.bpp.0 is written in the same format to record the initial parameter values used. This is done to avoid cluttering the <casename>.bpr file with what is often a very long list of parameters and their values. Parameter values that were subjected to logarithmic or power transformation are reported in their linear space, *not* log-transformed or power-transformed space.

Another special case of parameter value files is written at the end of a bgaPEST run and called <casename>.bpp.fin. This file contains the final parameter values estimated as optimal by bgaPEST. Furthermore, if posterior covariance calculation was requested, two additional columns are added: 95pctLCL and 95pctUCL, which are the 95 percent lower and upper confidence limits, respectively.

These confidence limits are obtained by applying the subtraction and addition, respectively, of $2 \times \sqrt{V_{ii}}$ to s_i—the ith optimal parameter value. In this case, $\mathbf{V}$ is the posterior covariance, so $\sqrt{V_{ii}}$ is the standard deviation of the ith parameter. The 95 percent confidence limits are reported in linear space, *not* log-transformed or power-transformed space, so for log-transformed or power-transformed parameters, the upper and lower 95 percent percent confidence limits are not symmetrical about the parameter value.

Observation Value Files

The observation values obtained by running the forward model with the currently estimated parameters are written to files called <casename>.bre.<#Oi>_<#Ii> following a similar convention as with the <casename>.bpp files above. The ASCII files are printed with the following headers: ObsName; ObsGroup; Modeled; Measured. These files can be easily copied into a spreadsheet or read with a plotting program to calculate and plot residuals.

Posterior Covariance File

If the input variable posterior_cov_flag=1, then posterior covariance of the parameters s is calculated. In addition to this information being used to report 95 percent confidence limits as described above, the posterior covariance matrix is also written to the file <casename>.post.cov. If the variable Q_compression_flag=1, then compression is used for saving the prior covariance matrix. This is done when many parameters are used and, thus, the full covariance matrices are unwieldy. On the basis of this choice, the posterior covariance is reported either as the diagonal of the posterior covariance matrix $(diag(\mathbf{V}))$ if Q_compression_flag=1 or the full covariance matrix $\mathbf{V}$ if Q_compression_flag=0. The output formats are discussed at the end of appendix 1.

Posterior Covariance and Parameter Transformations

In this section, it was indicated that in the <casename>.bpp.fin file, parameter values and 95 percent confidence intervals are reported in linear (untransformed) space, whereas in the <casename>.post.cov file, posterior covariance values are reported in estimation (log-transformed or power-transformed) space. Why the difference? The two files serve slightly different purposes. The parameter output file presents values in the units they are entered in (and, presumably, the units "seen" by the forward model). As a result, 95 percent confidence intervals are reported in the same way. Furthermore, the addition and subtraction of $2 \times \sqrt{V_{ii}}$ must be applied to the parameters before back-transformation, which explains the asymmetry of the confidence limits. On the other hand, the full posterior covariance matrix is intended for other analysis (propagation of variance through to predictions, conditional realizations, and others) in which the information should be retained in estimation (log-transformed or power-transformed) space. In the end, the decision of how to report these values is one of convention, and this side box is intended to make clear which was chosen in each case.

Suggestions and Guidelines for Initial Use

The Bayesian Geostatistical Approach is a highly parameterized method that is appropriate for some, but not all applications. In this section, we outline a few considerations to aid in the decision about whether to use bgaPEST on a given problem given the history and characteristics of the method. We also offer a few guidelines to help users avoid potential pitfalls in the application of bgaPEST.

This report documents the first release of bgaPEST and, to our knowledge, the first implementation of BGA available in a generalized package. As a result, users of this version will be among the first to apply this software outside of academia where custom programs have been the rule.

Nonetheless, the method has a 20-year history. The majority of applications have been to groundwater modeling projects including (but not limited to): pumping test analysis (Snodgrass and Kitanidis, 1998); hydraulic tomography (Li and others, 2007; Fienen and others, 2008; Li and others, 2008; Cardiff and Kitanidis, 2009; Cardiff and others, 2012); borehole logging (Fienen and others, 2004); contaminant source identification (Snodgrass and Kitanidis, 1997; Michalak and Kitanidis, 2002, 2003); and nonparametric tracer test analysis (Fienen and others, 2006). The main application to date that does not involve groundwater is in atmospheric modeling (Michalak and others, 2004; Mueller and others, 2008).

Characteristics of Appropriate bgaPEST Uses

The characteristics that unite these applications form a solid guide when deciding whether bgaPEST is appropriate for a given application. First and foremost, there must be a parameter set that varies in either space or time; for example, a time series of chemical concentrations (a breakthrough curve), a hydraulic conductivity field, a recharge field, or surface flux of atmospheric gases. These parameters should vary continuously over reasonably substantial areas so that a variogram serves as an adequate descriptor of the shape of the parameter field. Subareas delineated by geologic contacts—or in the case of time series, punctuated by known events—can be partitioned into beta associations, as discussed throughout this report. Another consideration is a more practical one: model run time.

The nature of bgaPEST is that many parameters are to be estimated. Throughout the parameter estimation process, a Jacobian sensitivity matrix must be calculated, requiring one model run per parameter. This computational burden must be considered and, potentially mitigated. In academic settings, many researchers have taken advantage of adjoint-state techniques to make the calculation of the Jacobian matrix more efficient in the case where parameters greatly outnumber observations. Adjoint-state versions of commercial and government codes are not typically available, however, but bgaPEST is

equipped to handle Jacobian matrices calculated outside of bgaPEST so that users who are able to write such codes can make use of them. Similarly, parallelization is supported to a limited degree using Python scripts and Condor (Condor Team, 2012) for run management. This parallel implementation is documented in appendix 4.

Adjoint-state Jacobian calculation is an attractive method to mitigate high computational expense of this method; however, production codes for adjoint state calculations are rare. For more information on the technique, see Townley and Wilson (1985), Sykes and others (1985), Samper and Neuman (1986), RamaRao and others (1995), and Neupauer and Wilson (1999) and references therein.

A common occurrence in groundwater modeling applications is that parameters far exceed observations in number. This, of course, can change in transient simulations where, if each measurement in time at a single measurement location is considered an observation, the numbers of observations and parameters may equalize. Use of bgaPEST is most appropriate for the former case—where parameters outnumber observations, typically by a large margin. Several programming and mathematical accommodations are made to enable the number of parameters to grow very large (testing has been performed with 90,000 parameters). If the number of observation grows significantly, however, computer memory will become a limitation in many cases. For transient problems, one should consider the information content of each measurement point in time. Often, the number of observation points can be effectively reduced by considering moments rather than discrete points (Li and others, 2005) or by other time-series processing such as methods available in R (R Development Core Team, 2011) or TSPROC (Westenbroek and others, 2012).

Guidelines

The number of applications of bgaPEST thus far is limited. Because bgaPEST is new software implementing a relatively novel technique, it will take time for users to get a feel for the behavior and characteristics of the tool. In this section, we provide a few guidelines that we hope will help users avoid pitfalls. In future releases, building on the experience of a larger user base, more guidelines will be available.

Run Times   For a typical groundwater model, somewhere between 5 and 15 outer iterations will often be required. For each outer iteration, it is likely that about 5 inner iterations will be necessary. This means as many as 75 calculations of the Jacobian matrix may be required. Without parallelization or adjoint state, users should carefully consider how many parameters can be accommodated as run times grow in length. For planning, assume that the time required for each Jacobian calculation will be the number of parameters $(m) \times$ Run Time.

Beta Associations   Beta associations provide the ability to include knowledge about contacts and other partitions in the parameter fields. Some beta associations can have a separate parameter value in each node, whereas others can be treated as homogeneous zones. Specification of either alternative is accomplished through the design of the template file. In addition to allowing for the inclusion of well-known structures such as lithologic contacts, beta associations also allow for some regions—either because of greater importance to the ultimate management decisions, or because of greater density of data, or both—to have a large number of parameters whereas other regions have homogeneous values. By allowing a large number of parameters only in focused areas of interest, the overall number of parameters can be reduced, thus mitigating some of the concerns about run times.

Line Search   The purpose of the line search is similar to the purpose of the Levenberg-Marquardt adjustment used in PEST. Whereas the Levenberg-Marquardt search makes a correction to the search direction when the optimization algorithm might otherwise stray from the optimal direction, the line search adjusts the length along the Quasi-Newton direction to avoid overshooting. The line search, therefore, serves its greatest purpose in its first iteration or two. After that, the value of the line search is limited for mathematical reasons having to do with linearization of the problem (see appendix 3 for more details). As a result, a value of between 2 and 5 for `it_max_linesearch` in the control variables is generally adequate. If the line search algorithm does not converge, a warning will be issued and, although it is good to know that this took place, the line search has served its purpose and little gain will be achieved by increasing `it_max_linesearch`.

Level of Fit   "With great power comes great responsibility." In applications where parameters outnumber observations, there lurks a real danger of overfitting. In other words, parameters can be adjusted to achieve of correspondence between simulated and observed equivalents that exceeds a reasonable level. The danger of this is that some of the *lack* of correspondence is often due to random epistemic error and is not representative of actual system behavior. However, if the parameters are adjusted to match observations within this margin of error, they are "fitting the noise." The ramifications of this type of adjustment are mainly diminished predictive power of the model and unrealistic roughness of the parameter fields estimated. There are two means of avoiding this problem. One is the maximum entropy property of BGA. The algorithm is designed to find the *smoothest* solution consistent with the level of fit. If all structural parameters—including σ_R—are estimated, then the algorithm will try to achieve perfect fit with the smoothest solution that can do so. This may still lead to overfitting, however, so in most cases, it is more appropriate to set the level of fit by using `sig_0` in the Epistemic Error Term input block described below to a level of fit chosen by the user to be appropriate given known and suspected uncertainties about both the observation quality and the model. Weights on observations can account for different levels of quality in different observations. In most cases, the user should also set `sig_opt=0` to force the algorithm to use a consistent value for epistemic uncertainty and thus manually control the level of fit. If set this way, the algorithm will adjust the other structural parameters to achieve the smoothest possible solution corresponding to the specified level of fit.

The level of smoothness in the optimal BGA solution is always smoother than conditional realizations (Kitanidis, 1995), which characterize more of the potential variability in each solution. In cases such as transport models where heterogeneity is the most important, conditional realizations (made possible using the optimal solution and the posterior covariance—both provided by bgaPEST) will result in a more precise characterization of system behavior in heterogeneity.

Structural Parameter Optimization   A good general guideline for all modeling is to start simple and add complexity as appropriate. In bgaPEST, this goal is achieved by starting with small values of variogram parameters (slope for the linear, variance for the nugget or exponential) such that the solution will be very smooth. By optimizing for structural parameters, roughness will be introduced by the algorithm until convergence at the optimal level of roughness is achieved. At the early, exploratory stages of a project, it might be desirable to set sig_opt=1 to see what level of fit may be achievable, but the user should be prepared to override this setting in later stages as allowing too much roughness to be introduced. For the prior distribution variogram parameters, however, optimization should always be employed in keeping with the Empirical Bayes perspective designed into the algorithm.

Limitations of bgaPEST Version 1.0

bgaPEST marks the first widely available implementation of BGA for use by practitioners. Limitations, of course, accompany this first implementation. For example, version 1.0 has a limited explicit parallelization facility. This can be overcome by using external programs for derivatives and calling a parallel Jacobian calculation package such a BeoPEST (Schreüder, 2009) or GENIE (Muffels and others, 2012) whenever a Jacobian matrix is required. The impact of this workaround is on the run times required to obtain a solution.

A practical upper limit on the number of parameters estimated is on the order of 100,000. To estimate a larger number of parameters, machines with a large amount of random access memory (RAM) must be used. At some greater limit, methods such as periodic embedding or other decompositions must be incorporated to mitigate the expense of storing and calculating the prior covariance matrix.

The source code is written in Fortran 90 and should be compilable on any platform with a Fortran

compiler. Special care was taken to avoid obscure and nonstandard language features. Nonetheless, it is possible that some platform- or compiler-specific problems may be encountered.

It is possible to use bgaPEST with a small number of parameters, but the assumption from the start is that parameters in at least part of the spatio-temporal domain represent a field of correlated instances (for example, model nodes or discrete times) that often outnumber the number of data observations. A combination of homogeneous parameters in zones with a refined area of interest that is distributed is a common application and, as implemented through beta associations, this mix of distributed and zoned parameters is supported and encouraged. Typically, sufficient data to support a distributed parameter set are limited to part of a model domain in space or time.

In considering uncertainty, version 1.0 presents posterior covariance values. For some applications, conditional realizations may be desired to capture candidate roughness of solutions within the ensemble distribution of solutions. Details for conditional realizations are provided by Kitanidis (1995).

Acknowledgments

This work was supported by a variety of projects and programs. General support was provided by the National Research Council Postdoctoral Research Associateship, the U.S. Geological Survey (USGS) Groundwater Resources Program, and the USGS Water Energy and Biogeochemical Budgets (WEBB) Program. In addition, the USGS Exchange Visitor Program and the USGS Wisconsin Water Science Center supported a 6-month exchange in which Dr. Marco D'Oria was hosted in Wisconsin. Dr. D'Oria also received support from the University of Parma, Italy, Department of Civil and Environmental Engineering and Architecture (Università degli Studi di Parma). The authors thank Dr. Michael Cardiff, University of Wisconsin—Madison, Stephen Westenbroek, USGS Wisconsin Water Science Center, and Mike Eberle, USGS Science Publishing Network, for valuable technical and editorial reviews of this document. The authors finally thank Dr. Peter Kitanidis, Stanford University, for his pioneering work on this technique and mentoring the first author as a graduate advisor.

References Cited

Aster, R.C., Borchers, B., and Thurber, C.H., 2005, Parameter estimation and inverse problems: Amsterdam, Elsevier Academic Press, International Geophysics Series, v. 90, 301 p.

Banta, E.R., Poeter, E.P., Doherty, J.E., and Hill, M.C., 2006, JUPITER: Joint Universal Parameter IdenTification and Evaluation of Reliability—An application programming interface (API) for model analysis: U.S. Geological Survey Techniques and Methods, book 6, chap. E1, 268 p.

Cardiff, M., and Kitanidis, P.K., 2009, Bayesian inversion for facies detection—An extensible level set framework: Water Resources Research, v. 45, W10416, doi:10.1029/2008wr007675.

Cardiff, M., Barrash, W., and Kitanidis, P.K., 2012, A field proof-of-concept of aquifer imaging using 3-D transient hydraulic tomography with modular, temporarily-emplaced equipment: Water Resources Research, v. 48, no. 5, W05531, doi:10.1029/2011WR011704.

Casella, G., 1985, An introduction to empirical Bayes data-analysis: American Statistician, v. 39, no. 2, p. 83–87, doi:10.2307/2682801.

Chamberlin, T.C., 1890, The method of multiple working hypotheses: Science (Old Series), v. 15, no. 92.

Collinson, J.D., 1969, Sedimentology of Grindslow shales and Kinderscout grit—A deltaic complex in Namurian of Northern England: Journal of Sedimentary Petrology, v. 39, no. 1, p. 194–221.

Condor Team, 2012, Condor Version 7.6.6 Manual: Madison, Wisconsin, University of Wisconsin—Madison.

Deutsch, C.V., and Journel, A.G., 1992, GSLIB—Geostatistical software library and users guide: New York, Oxford University Press, 340 p.

Doherty, J., 2010a, PEST, Model-independent parameter estimation—User manual (5th ed., with slight additions): Brisbane, Australia, Watermark Numerical Computing.

Doherty, J., 2010b, PEST, Model-independent parameter estimation—Addendum to user manual (5th ed.): Brisbane, Australia, Watermark Numerical Computing.

Draper, N.R., and Smith, H., 1966, Applied regression analysis: New York, Wiley, 407 p.

Fienen, M., Kitanidis, P., Watson, D., and Jardine, P., 2004, An application of Bayesian inverse methods to vertical deconvolution of hydraulic conductivity in a heterogeneous aquifer at Oak Ridge National Laboratory: Mathematical Geology, v. 36, no. 1, p. 101–126, doi:10.1023/B:MATG.0000016232.71993.bd.

Fienen, M., Luo, J., and Kitanidis, P., 2006, A Bayesian geostatistical transfer function approach to tracer test analysis: Water Resources Research, v. 42, no. 7, W07426, doi:10.1029/2005WR004576.

Fienen, M.N., Clemo, T.M., and Kitanidis, P.K., 2008, An interactive Bayesian geostatistical inverse protocol for hydraulic tomography: Water Resources Research, v. 44, W00B01, doi:10.1029/2007WR006730.

Fienen, M., Hunt, R., Krabbenhoft, D., and Clemo, T., 2009, Obtaining parsimonious hydraulic conductivity fields using head and transport observations—A Bayesian geostatistical parameter estimation approach: Water Resources Research, v. 45, W08405, doi:10.1029/2008wr007431.

Hill, M.C., 2006, The practical use of simplicity in developing ground water models: Ground Water, v. 44, no. 6, p. 775–781, doi:10.1111/j.1745-6584.2006.00227.x.

Hoeksema, R.J., and Kitanidis, P.K., 1984, An application of the geostatistical approach to the inverse problem in two-dimensional groundwater modeling: Water Resources Research, v. 20, no. 7, p. 1003–1020, doi:10.1029/WR020i007p01003.

Isaaks, E.H., and Srivastava, R.M., 1989, Applied geostatistics: Oxford, UK; New York; Oxford University Press, 561 p.

Jaynes, E.T., and Bretthorst, G.L., 2003, Probability theory—The logic of science: Cambridge, UK; New York; Cambridge University Press, 727 p.

Kitanidis, P.K., 1995, Quasi-linear geostatistical theory for inversing: Water Resources Research, v.

31, no. 10, p. 2411–2419, doi:10.1029/95WR01945.

Kitanidis, P.K., 1997, Introduction to geostatistics—Applications in hydrogeology: Cambridge, UK; New York; Cambridge University Press, 249 p.

Kitanidis, P.K., and Vomvoris, E.G., 1983, A geostatistical approach to the inverse problem in groundwater modeling (steady state) and one-dimensional simulations: Water Resources Research, v. 19, no. 3, p. 677–690, doi:10.1029/WR019i003p00677.

Li, W., Nowak, W., and Cirpka, O.A., 2005, Geostatistical inverse modeling of transient pumping tests using temporal moments of drawdown: Water Resources Research, v. 41, no. 8, p. 1–13, doi:10.1029/2004WR003874.

Li, W., Englert, A., Cirpka, O.A., Vanderborght, J., and Vereecken, H., 2007, Two-dimensional characterization of hydraulic heterogeneity by multiple pumping tests: Water Resources Research, v. 43, no. 4, W04433, doi:10.1029/2006WR005333.

Li, W., Englert, A., Cirpka, O.A., and Vereecken, H., 2008, Three-dimensional geostatistical inversion of flowmeter and pumping test data: Ground Water, v. 46, no. 2, p. 193–201, doi:10.1111/j.1745-6584.2007.00419.x

Michalak, A.M., Bruhwiler, L., and Tans, P.P., 2004, A geostatistical approach to surface flux estimation of atmospheric trace gases: Journal of Geophysical Research, v. 109, no. D14, doi:10.1029/2003jd004422.

Michalak, A.M., and Kitanidis, P.K., 2002, Application of Bayesian inference methods to inverse modeling for contaminant source identification at Gloucester Landfill, Canada, *in* Hassanizadeh, S.M., Schotting, R.J., Gray, W.G., and Pinder, G.F., eds., Proceedings of the Fourteenth International Conference on Computational Methods in Water Resources (CMWR XIV): Amsterdam, Elsevier, v. 2, p. 1259–1266.

Michalak, A.M., and Kitanidis, P.K., 2003, A method for enforcing parameter nonnegativity in Bayesian inverse problems with an application to

contaminant source identification: Water Resources Research, v. 39, no. 2, 1033, doi:10.1029/2002WR001480.

Mueller, K.L., Gourdji, S.M., and Michalak, A.M., 2008, Global monthly averaged CO_2 fluxes recovered using a geostatistical inverse modeling approach; 1. Results using atmospheric measurements: Journal of Geophysical Research-Atmospheres, v. 113, no. D21, doi:10.1029/2007jd009734.

Muffels, C., Schreüder, W., Doherty, J., Karanovic, M., Tonkin, M., Hunt, R., and Welter, D., 2012, Approaches in highly parameterized inversion—GENIE, a general model-independent TCP/IP run manager, U.S. Geological Survey Techniques and Methods, book 7, chap. C6, 26 p.

Neupauer, R.M., and Wilson, J.L., 1999, Adjoint method for obtaining backward-in-time location and travel time probabilities of a conservative groundwater contaminant: Water Resources Research, v. 35, no. 11, p. 3389–3398.

Nowak, W., and Cirpka, O.A., 2004, A modified Levenberg-Marquardt algorithm for quasi-linear geostatistical inversing: Advances in Water Resources, v. 27, no. 7, p. 737–750, doi:10.1016/j.advwatres.2004.03.004.

R Development Core Team, 2011, R—A language and environment for statistical computing: Vienna, Austria, R Foundation for Statistical Computing, ISBN 3-900051-07-0.

RamaRao, B.S., Lavenue, A.M., de Marsily, G., and Marietta, M.G., 1995, Pilot point methodology for automated calibration of an ensemble of conditionally simulated transmissivity fields; 1. Theory and computational experiments: Water Resources Research, v. 31, no. 3, p. 475–493.

Remy, N., Boucher, A., and Wu, J., 2009, Applied geostatistics with SGeMS: Cambridge, UK; New York; Cambridge University Press, 264 p.

Robbins, H., 1956, An empirical Bayes approach to statistics, *in* Neyman, J., ed., Proceedings of the Third Berkeley Symposium on Mathematical Statistics: University of California Press, v. 1, p. 157–163.

Rubin, Y., 2003, Applied stochastic hydrogeology: Oxford, UK; New York; Oxford University Press,

391 p.

Samper, F.J., and Neuman, S., 1986, Adjoint state equations for advective-dispersive transport, *in* Sixth International Conference on Finite Elements in Water Resources, p. 423–437.

Schreüder, W., 2009, Running BeoPEST, *in* Proceedings, PEST Conference 2009, Potomac, Md., November 1–3, 2009: Bethesda, Md., S.S. Papadopulos and Associates, p. 228–240.

Snodgrass, M.F., and Kitanidis, P.K., 1997, A geostatistical approach to contaminant source identification: Water Resources Research, v. 33, no. 4, p. 537–546.

Snodgrass, M., and Kitanidis, P., 1998, Transmissivity identification through multi-directional aquifer stimulation: Stochastic Hydrology and Hydraulics, v. 12, no. 5, p. 299–316, doi:10.1007/s004770050023.

Swift, D.J.P., Parsons, B.S., Foyle, A., and Oertel, G.F., 2003, Between beds and sequences—Stratigraphic organization at intermediate scales in the Quaternary of the Virginia coast, USA: Sedimentology, v. 50, no. 1, p. 81–111, doi:10.1046/j.1365-3091.2003.00540.x.

Sykes, J.F., Wilson, J.L., and Andrews, R.W., 1985, Sensitivity analysis for steady state groundwater flow using adjoint operators: Water Resources Research, v. 21, no. 3, p. 359–371, doi:10.1029/WR021i003p00359.

Tikhonov, A.N., 1963a, Solution of incorrectly formulated problems and the regularization method [in Russian]: Soviet Mathematics Doklady, v. 4, p. 1035–1038.

Tikhonov, A.N., 1963b, Regularization of incorrectly posed problems [in Russian]: Soviet Mathematics Doklady, v. 4, p. 1624–1637.

Townley, L., and Wilson, J., 1985, Computationally efficient algorithms for parameter estimation and uncertainty propagation in numerical models of groundwater flow: Water Resources Research, v. 21, no. 12, p. 1851–1860.

Walker, R.G., 1984, General introduction—Facies, facies sequences and facies models, *chap. 1 of* Walker, R.G., Facies models (2d ed.): Toronto, Geological Association of Canada, p. 1–9.

Walker, R.G., 1992, Facies, facies models and modern stratigraphic concepts, *chap. 1 of* Walker, R.G., and James, N.P., Facies models—Response to sea level change: St. Johns, Newfoundland, Geological Association of Canada, p. 1–14.

Westenbroek, S., Doherty, J., Walker, J., Kelson, V., Hunt, R., and Cera, T., 2012, Approaches in highly parameterized inversion—TSPROC, a general time-series processor to assist in model calibration and result summarization: U.S. Geological Survey Techniques and Methods, book 7, chap. C7, 79 p.

Appendix 1—Input Instructions

In this appendix, the general strategy for input instructions is described. The input is arranged in a file called <casename>.bgp, which is made up of input blocks, as discussed below. Following a discussion of more detail of the general input protocols, subsections are presented in which specific input blocks are discussed, including variables and data that can be inserted.

General Structure of Input

The general input structure is designed on a subset of the JUPITER protocol (Banta and others, 2006). The advantage of this protocol over XML or the previous input format for PEST is that annotations that are easily read by humans are part of the input protocol. The full JUPITER protocol, however, has memory and computational overhead that can become a problem for large and complicated datasets. The protocol used here, therefore, is simplified but should be easily recognizable to users of other JUPITER-compatible programs.

The strategy for input is designed to use BLOCKS that are made up of either KEYWORDS for individual variables or TABLES for a series of data. The specification of whether a given block uses KEYWORDS or TABLES is predetermined and the input blocks defined below indicate which is required.

Blocks

Input blocks are allowed to take one of two forms: either KEYWORDS or TABLES. All input blocks are delineated by the words BEGIN and END. The header line also includes the name and type of the block and the final line contains the name of the block. For example,

```
BEGIN prior_mean_cv KEYWORDS
prior_betas=1
beta_cov_form = 0
END prior_mean_cv
```

Keywords

Keyword variables correspond to single values identified with an "=" sign. Multiple KEYWORDS can be entered on each line in an input file, but no spaces are allowed in KEYWORDS names or variable values. An example is: prior_betas=1.

Tables

Table variables are used for tabular data series that have multiple values in categories. Tables are identified by listing the number of rows (nrow), and number of columns (ncol), and by providing the keyword columnlabels. This is followed by nrow rows of data, with values arranged in ncol columns, corresponding to the same order as the columnlabels and delimited by one or more spaces. For example,

```
BEGIN Q_compression_cv TABLE
nrow=2 ncol=5 columnlabels
BetaAssoc Toep_flag Nrow Ncol Nlay
1 0 21 21 1
2 1 21 21 1
END Q_compression_cv
```

Files

A user may want to shorten the length of the main input file by reading certain input from external text files. This can be done by signaling an input block with the word FILES, to read a file containing the entire set of information for the block. Regardless of whether the external text file contains a KEYWORDS or TABLE block, a block definition must be in place directing the program to the external file. For example,

```
BEGIN Q_compression_cv FILE
compression.txt
END Q_compression_cv
```

In this example, the contents of the file compression.txt would be

```
BEGIN Q_compression_cv TABLE
nrow=2 ncol=5 columnlabels
BetaAssoc Toep_flag Nrow Ncol Nlay
1 0 21 21 1
2 1 21 21 1
END Q_compression_cv
```

bgaPEST Input Blocks

The specific input blocks used in bgaPEST are discussed, in order of appearance in the <casename>.bgp file. It is important to maintain the order of the blocks in the same order as discussed in this report. For each block, data types are identified either as *float*, *integer*, or *string*. Values entered as *float* can include scientific/engineering notation, but in all cases should contain a "." even if no fractional detail is included. Conversely, *integers* must not contain ".". Variables identified as *string* may not include spaces because whitespace is used as the delimiter for rows in tables and separating keywords.

Each block is also defined with a suffix of "cv" for "control variables" or "data" for data. Control variables are those that govern the behavior of the algorithm as a whole as opposed to data points (such as parameter values, structural parameter values, and so forth).

> **A note on default variable values**
>
> In the input instructions below, some variables list a default value. Part of the design strategy of this software was to not burden users with determining appropriate values for each and every variable that controls the algorithm. As a result, default values are provided for some variables. In those cases, input by the user in the .bgp file is optional. If no value is provided by the user, the default value will be used by bgaPEST. If a variable not listed with a default value in these input instructions is omitted by a user, bgaPEST will return with an error indicating that the variable is not present.

Algorithmic Control Variables (algorithmic_cv) KEYWORDS

The following KEYWORDS variables are in the algorithmic_cv block.

structural_conv *float,default=0.001* Convergence criterion for structural parameter convergence. If positive, convergence is based on the absolute difference in structural parameter objective function over consecutive iterations. If negative, convergence is based on the norm of the difference between consecutive structural parameter values. Used only if at least one structural parameter is to be estimated.

phi_conv *float, default=0.001* Convergence criterion for objective function inner iterations.

bga_conv *float, default=10×phi_conv* Convergence criterion for objective function outer iterations.

it_max_structural *integer, default=10* Total number of iterations allowed in structural parameter optimization.

it_max_phi *integer, default=10* Total number of iterations allowed in each quasi-linear estimation optimization.

it_max_bga *integer, default=10* Total number of outer iterations allowed for the entire algorithm.

linesearch *integer, default=0* Flag to determine whether a line search should be conducted. [0] = do not use line search, [1] = use line search.

it_max_linesearch *integer, default=4* Total number of outer iterations allowed for the line search. Used only if linesearch = 1.

theta_cov_form *integer, default=0* Form of the theta covariance matrix. [0] = none, [1] = diagonal, [2] = full matrix. [0] means no prior covariance on theta provided and it is assumed to be unknown. Used only if at least one structural parameter is to be estimated.

Q_compression_flag *integer, default=0* Flag to determine how to calculate $\mathbf{Q}_{ss}$. [0] = no compression—calculate full $\mathbf{Q}_{ss}$ matrix, [1] = Calculate separate $\mathbf{Q}_{ss}$ matrix for each beta association.

par_anisotropy *integer, default=0* Flag to determine whether parameter anisotropy should be considered when making the $\mathbf{Q}_{ss}$ matrix. [0] = do not consider anisotropy, [1]=consider anisotropy. If anisotropy is considered, a parameter_anisotropy block should be included, as defined below.

deriv_mode *integer, default=0* Flag to determine whether sensitivities are calculated by using an external call to PEST or using a user-supplied program (such as adjoint state). [0] = use PEST, [1] = use external program identified below in the model_command_lines block, [4] = use external derivatives in parallel (see appendix 4 for details).

posterior_cov_flag *integer, default=0* Flag to determine whether posterior covariance matrix should be calculated. [0] = do not calculate posterior covariance matrix, [1] = calculate posterior covariance matrix. If Q_compression_flag = 1, only the diagonal of the posterior covariance matrix is calculated.

jacobian_file *string, default="scratch.jco"* Name of the file generated by an external program if deriv_mode = 1. If deriv_mode = 0, this value is ignored and left at its default value.

jacobian_format *string, default="binary"* Format of the file indicated in jacobian_file. [binary] indicates a binary file formatted as a JCO file from PEST, [ascii] indicates a file of a standard PEST matrix format, discussed below in this documentation. If deriv_mode = 0, this value is ignored and left at its default value.

Prior Mean Control Variables (prior_mean_cv) KEYWORDS

The following KEYWORDS variables are in the prior_mean_cv block.

prior_betas *integer* Flag indicating whether information about prior mean (β) will be supplied. [0] = no, [1] = yes.

beta_cov_form *integer, default=0* Form of the prior mean (β) covariance matrix $\left(\mathbf{Q}_{\beta\beta}\right)$. [0] = none, [1] = diagonal, [2] = full matrix. This value is used only if prior_betas = 1.

Beta Association Data (prior_mean_data) TABLE

This table must contain the same number of rows as there are beta associations to be defined. The rows must be in ascending order of beta association numbers. This is also the block where beta associations are defined, even if prior means are not defined. Parameter transformations are also defined in this table. Details about transformation options are in appendix 3.

BetaAssoc *integer* Identifier of each beta association (one per row). These should be sequential integers.

Partrans *string* Transformation indicator determining whether β values will be in physical or estimation space. Acceptable values are log, power, and none.

alpha_trans *float, default = 50.0* Exponent of the power transformation if Partrans = power.

beta_0 *float* Value of prior mean (β_0) for the row's beta association. This value is used only if prior_betas = 1.

beta_cov_# *float* The number of values provided is based on the value of beta_cov_form specified above:

> If beta_cov_form = 1, one value is provided.
> If beta_cov_form = 2, nrow values are provided, corresponding to the current row of the beta covariance matrix $(\mathbf{Q}_{\beta\beta})$.
> This value is used only if prior_betas = 1.

Structural Parameter Control Variables (structural_parameter_cv) TABLE

This table must contain the same number of rows as there are beta associations to be defined. The rows must be in ascending order of beta association numbers.

BetaAssoc *integer* Identified for each beta association (one per row). These should be sequential integers.

prior_cov_mode *integer, default = 1* Flag to indicate whether prior covariance of parameters ($\mathbf{Q}_{ss}$) is supplied or calculated. This is reserved for future use. Currently, $\mathbf{Q}_{ss}$ is always calculated, so this value is ignored if present.

var_type *integer, default=1* This is a flag to indicate which variogram type is used to express the prior covariance ($\mathbf{Q}_{ss}$). Acceptable choices are [0] = pure nugget, [1] = linear, [2] = exponential.

struct_par_opt *integer, default=1* Flag for whether structural parameters are meant to be optimized or not. This can be chosen for each structural parameter individually, [0] = do not optimize (hold at initial value), [1] = optimize by using a marginal distribution.

trans_theta *integer, default=0* Flag for whether a power transformation should be applied to the structural parameters in the current row. [0] = do not transform, [1] = transform. This value is used only if struct_par_opt = 1.

alpha_trans *float, default = 50* Exponent of the power transformation, used only if trans_theta = 1. Details of the power transformation are in appendix 3.

Structural Parameter Data (structural_parameter_data) TABLE

This table must contain the same number of rows as there are beta associations to be defined. The rows must be in ascending order of beta association numbers.

BetaAssoc *integer* Identifier of each beta association (one per row). These should be sequential integers.

theta_0_1 *float* Initial value of $\theta_{1,0}$, which is the starting value of the first structural parameter for prior covariance.

theta_0_2 *float* Initial value of $\theta_{2,0}$, which is the starting value of the second structural parameter for prior covariance. If a linear or nugget variogram is used, an arbitrary negative value should be entered here indicating that the value will be ignored. For an exponential variogram, this parameter is the correlation length.

Structural Parameter Covariance Data (structural_parameter_cov) TABLE

The only covariance model currently supported is diagonal, so there must be one covariance value for each θ parameter. This block is read only if theta_cov_form is not zero.

theta_cov_1 *float* Variance of the current row's θ parameter. If an exponential variogram is used, then a single beta association will have two structural parameters. To handle this possibility, the variance values should be listed, one per line, in the order of beta associations, then in order θ_1 then θ_2. Even structural parameters that will not be estimated (for example, that are held at their initial values, as indicated by struct_par_opt above) must have a placeholder value entered here to maintain the order—the placeholder value is arbitrary and will be ignored.

Epistemic Error Term (epistemic_error_term) KEYWORDS

sig_0 *float* Initial value of the epistemic uncertainty variance $\left(\sigma_{R_0}^2\right)$.

sig_opt *integer* Flag indicating whether epistemic uncertainty variance should be optimized for or not. [0] = do not optimize, [1] = optimize. If sig_opt = 0, then the value of sig_0 is used throughout the inversion.

sig_p_var *float, default=0* Prior variance on $\sigma_{R_0}^2$. sig_p_var = 0 means no prior variance on epistemic error is provided and it is assumed totally unknown. This value is used only if sig_opt = 1.

trans_sig *integer, default=0* Flag for whether a power transformation should be applied to the epistemic error. [0] = do not transform, [1] = transform. This value is used only if sig_opt = 1.

alpha_trans *float, default = 50* Exponent of the power transformation, used only if trans_sig = 1. Details of the power transformation are in appendix 3.

Parameter Control Variables (parameter_cv) KEYWORDS

ndim *integer* Number of dimensions over which the estimated parameters span.

Prior Covariance Compression Control Variables (Q_compression_cv) TABLE

This table must contain the same number of rows as there are beta associations to be defined. The rows must be in ascending order of beta association numbers. This block is read only if Q_compression_flag = 1.

BetaAssoc *integer* Identified of each beta association (one per row). These typically are sequential integers.

Toep_flag *integer* This is a flag to determine whether a Toeplitz transformation should be applied to the prior covariance matrix ($\mathbf{Q_{ss}}$). [0] = do not use Toeplitz transformation, [1] = use Toeplitz transformation.

Nrow *integer* Number of rows in the current beta association (read only if Toep_flag = 1).

Ncol *integer* Number of columns in the current beta association (read only if Toep_flag = 1).

Nlay *integer* Number of layers in the current beta association (read only if Toep_flag = 1).

Parameter Groups (`parameter_groups`) TABLE

Each row of this table corresponds to one of the parameter groups. These groups are used to group together parameters by type and are not the same as beta associations.

`groupname` *string* Name of the group in the current row. Note that these cannot contain spaces.

`grouptype` *integer* Integer identifying which type of parameter the group corresponds to. This is used to ensure that beta associations do not span parameter types (for example, hydraulic conductivity parameters should not be in the same group type as recharge parameters). The specific values are arbitrary, but a distinct value should be assigned to each parameter group type.

`derinc` *float* The derivative increment for calculating external derivates if using external derivatives calculation (`deriv_mode = 0` or `deriv_mode = 4`).

Parameter Data (`parameter_data`) TABLE

Each row of this table provides information for one parameter.

`ParamName` *string* Name for the parameter.

`StartValue` *float* Starting parameter value.

`GroupName` *string* Name of the group to which the parameter belongs. This name must be defined in the `parameter_groups` block.

`BetaAssoc` *integer* Beta association to which this parameter belongs.

`SenMethod` *integer* Sensitivity method used for this parameter type. This parameter may now be arbitrary—it is reserved for future use and currently is ignored.

`x1` *float* Location in the first dimension.

`x2` *float* Location in the second dimension. Read only if `ndim` $>=$ 2.

`x3` *float* Location in the third dimension. Read only if `ndim` = 3.

Observation Groups (`observation_groups`) TABLE

Each row of this table corresponds to one of the observation groups. These groups are used to group together observations by type and are used to report portions of the objective function.

`groupname` *string* Name of the group in the current row. Note that these cannot contain spaces.

Observation Data (`observation_data`) TABLE

One observation is presented on each line.

`ObsName` *string* Name of an observation.

`ObsValue` *float* Value of the observation.

`GroupName` *string* Name of the group to which the observation belongs. This name must be defined in the `observation_groups` block.

`Weight` *float* A relative weight that gets applied to the epistemic error.

Model Command Lines (`model_command_lines`) KEYWORDS

Currently, a single forward model command and an optional derivative model command can be supplied here. These string keywords can include path information if the command line batch files or shell scripts are not located in the current working directory, but spaces are not allowed.

`Command` *string* This is the batch file or shell script that runs the forward model.

`DerivCommand` *string* This is the optional batch file or shell script that is used to calculate derivatives. This is used only if `deriv_method = 1` in the `algorithmic_cv` block.

Model Input Files (`model_input_files`) TABLE

Each row of this table includes a matched template file and model input file. This allows the program to create the correct input files for the model.

`TemplateFile` *string* Name of a template file for making model input. Must end in `.tpl`.

`ModInFile` *string* Name of the model input file corresponding to the TemplateFile identified on the same row.

Model Output Files (`model_output_files`) TABLE

Each row of this table includes a matched instruction file and model output file. This allows the program to read the results of model runs correctly.

`InstructionFile` *string* Name of an instruction file for reading model output. Must end in `.ins`.

`ModOutFile` *string* Name of the model output file corresponding to the InstructionFile identified on the same row.

Parameter Anisotropy (`parameter_anisotropy`) TABLE

Each row of this table contains information for parameter anisotropy for a beta association. This block is read only if the variable `par_anisotropy` =1 in the `algorithmic_cv` block.

`BetaAssoc` *integer* Identifier of a beta association.

`horiz_angle` *float* Angle, in degrees of the principal direction of anisotropy in a horizontal plane. See figure 3.1 for details.

`horiz_ratio` *float* Ratio of maximum to minimum principal property values in the horizontal plane. See figure 3.1 for details.

`vertical_ratio` *float* Ratio of maximum to minimum principal property values in the vertical direction. See figure 3.1 for details. This value is read only if `ndim=3`.

PEST Matrix Formats for Jacobian and Posterior Covariance

On two occasions in bgaPEST, a matrix text file format from PEST is used to store matrices: when posterior covariance output from bgaPEST is specified as a full matrix; and when Jacobian sensitivity matrix information is exchanged from an external code with bgaPEST run through, for example, a Python script.

```
    3    4    2
3.4423    23.323    2.3232    1.3232
5.4231    3.3124    4.4331    3.4442
7.4233    5.4432    7.5362    8.4232
* row names
apar1
apar2
apar3
* column names
aobs1
aobs2
aobs3
aobs4
```

Figure 1.1.   Example of a standard PEST matrix file, adapted from Doherty (2010b).

The posterior covariance matrix may take two forms: a full matrix or a diagonal matrix. These options are discussed below. Two options are available for Jacobian sensitivity matrices (**H**) to be read by bgaPEST. If deriv_mode=0, PEST is used, external to bgaPEST, to calculate the Jacobian matrix resulting in a binary file with the extension .jco. If deriv_mode=1 then an external program is used to calculate **H**, and a file, written by the external program, must be communicated to bgaPEST. This file can either be a .jco file or a .jac file which is an ASCII file following the format of a standard matrix file used by PEST, as described in Doherty (2010b), section 4.4.3. An example and description of the format of a standard PEST matrix follow, quoting from Doherty (2010b).

Figure 1.1 depicts an example matrix file holding a matrix with three rows and four columns.

The first line of a matrix file contains 3 integers. The first two indicate the number of rows (NROW) and number of columns (NCOL) in the following matrix. The next integer (named ICODE) is a code, the role of which will be discussed shortly. Following the header line is the matrix itself, in which entries are space-separated and wrapped to the next line if appropriate. The maximum line length is 500 characters, so wrapping to the next line must occur within 500 characters. It is recommended to wrap lines after 8 values and to maintain maximum possible precision.

In use for Jacobian matrices by bgaPEST, ICODE is set to 2, so the string "* row names" is printed next, followed by NROW names (of 20 characters or less in length), containing the names associated with rows of the matrix. NCOL column names follow in a similar format, following the string "* column names".

Other options for ICODE are described in Doherty (2010b) and are used in bgaPEST for output of the posterior covariance matrix. The two options for posterior covariance output both refer to square matrices that have the same names of columns and rows. As a result, only one list of names follows the data following the string "* row and column names".

If compression is used in the prior covariance matrix, bgaPEST outputs only the diagonal elements of the posterior covariance. In this case, ICODE=-1 and only the diagonal entries are listed, one per line, after the header line. If compression is not used, the entire posterior covariance matrix is printed using ICODE=1 with 8 values per line.

References Cited

Banta, E.R., Poeter, E.P., Doherty, J.E., and Hill, M.C., 2006, JUPITER: Joint Universal Parameter IdenTification and Evaluation of Reliability—An application programming interface (API) for model analysis: U.S. Geological Survey Techniques and Methods, book 6, chap. E1, 268 p.

Doherty, J., 2010b, PEST, Model-independent parameter estimation—Addendum to user manual (5th ed.): Brisbane, Australia, Watermark Numerical Computing.

Appendix 2—Quick Start Instructions

One advantage of using block input and keywords, as discussed in appendix 1, is that default values are supplied within bgaPEST so they can be skipped by a user. The values supplied as defaults have general applicability and will all be reported in the `<casename>.bpr` file. In this section, then, the bare minimum level of input is described to get a project running.

Forward Model

The forward model must exist and have the ability—either inherently or through pre- and post-processing—to receive input and provide output using text (ASCII) files. For bgaPEST to be able to run the model, template files (`.TPL`) and instruction files (`.INS`) must be provided, corresponding with model input and output, respectively. Details of the construction of these files are in Doherty (2010, chap. 3). The template and instruction files are detailed in the `model_input_files` and `model_output_files` blocks, respectively. The `model_command_lines` block also must be included with an entry for either a batch file or shell script in the `command` keyword that runs the model.

Observations

The `observation_groups` block must be completed. All observations can belong to the same group if desired. Groups are reported in output to assist in interpretation of results. The `observation_data` block also must be completed.

Beta Associations

Beta associations are first defined in the `prior_mean_data` block. If no prior information about mean values and their covariance is to be supplied, the only information necessary is a row for each beta association in the `prior_mean_data` block and a decision about whether to transform the value with a logarithmic or power transform. Note that beta associations indicate regions and groups that will have the same mean value estimated regardless of whether prior information about the mean is provided.

Structural Parameters

Each beta association must have a variogram specified for it, defined by structural parameters. Therefore, the `structural_parameters_cv` and `structural_parameter_data` blocks must be completed. Whether to optimize for structural parameter values and whether to provide prior information about the values are optional.

Parameters

The `parameter_groups` block must be completed and, like with observations, it is acceptable for all parameters to be in a single group, and groups do not need to correspond with beta associations. The `parameter_cv` keyword `ndim` must be provided, and the `parameter_data` block must be completed.

Algorithmic Control Variables

The `algorithmic_cv` block contains variables that all have default values; however, bgaPEST must find the `algorithmic_cv` block—even if it is empty. If the `algorithmic_cv` block is empty, all default values will be used.

Example .bgp Input File

The following text shows a bgaPEST input file. Two dependent files for parameters and observations are shown in abbreviated form to indicate their format.

Example1.bgp
```
BEGIN algorithmic_cv KEYWORDS
structural_conv=0.004 phi_conv=0.004
bga_conv=1.0e-2 it_max_structural=10
it_max_phi=15 it_max_bga=15
linesearch=1 it_max_linesearch=3
theta_cov_form=1 Q_compression_flag=1 deriv_mode = 1
jacobian_format = ascii jacobian_file = S1_1.jac
posterior_cov_flag = 1 par_anisotropy=0
END algorithmic_cv
BEGIN prior_mean_cv KEYWORDS
prior_betas= 1 beta_cov_form=1
END prior_mean_cv
BEGIN prior_mean_data TABLE
nrow=1 ncol=7 columnlabels
BetaAssoc Partrans beta_0 beta_cov_1 beta_cov_2 beta_cov_3 beta_cov_4
1 log -7.6 5e-7 65.   3.1 4.1
END prior_mean_data
BEGIN structural_parameter_cv TABLE
nrow=1 ncol=6 columnlabels
BetaAssoc prior_cov_mode var_type struct_par_opt trans_theta alpha_trans
1 2 1 1 1 20
END structural_parameter_cv
BEGIN structural_parameters_data TABLE
nrow=1 ncol=3 columnlabels
BetaAssoc theta_0_1 theta_0_2
1 1.0e-007 -0.1
END structural_parameters_data
BEGIN structural_parameters_cov TABLE
nrow=1 ncol=1 columnlabels
theta_cov_1
11.1
END structural_parameters_cov
BEGIN epistemic_error_term KEYWORDS
sig_0 = 1.000e-000 sig_opt = 0 sig_p_var=0.00001
END epistemic_error_term
BEGIN parameter_cv KEYWORDS
ndim=3
END parameter_cv
BEGIN Q_compression_cv TABLE
nrow=1 ncol=5 columnlabels
BetaAssoc Toep_flag Nrow Ncol Nlay
1 1 21 21 1
END Q_compression_cv
BEGIN parameter_groups TABLE
```

```
nrow=1 ncol=2 columnlabels
groupname grouptype
pargp_uno 1
END parameter_groups
BEGIN parameter_data FILES
PARAMETERS.txt
END parameter_data
BEGIN observation_groups TABLE
nrow=1 ncol=1 columnlabels
groupname obsgp_oden
END observation_groups
BEGIN observation_data FILES
obs12.txt
END observation_data
BEGIN model_command_lines KEYWORDS
Command = modflow.bat
DerivCommand = modflow_adj.bat
END model_command_lines
BEGIN model_input_files TABLE
nrow=1 ncol=2 columnlabels
TemplateFile ModInFile
S1_mul.tpl S1_.mul
END model_input_files
BEGIN model_output_files TABLE
nrow=1 ncol=2 columnlabels
InstructionFile ModOutFile
S1_1_hbs.ins S1_1.hbs
END model_output_files
BEGIN parameter_anisotropy TABLE
nrow = 1 ncol = 4 columnlabels
BetaAssoc horiz_angle horiz_ratio vertical_ratio
1 45 10 10
END parameter_anisotropy
```

PARAMETERS.txt

```
BEGIN parameter_data
TABLE nrow=441 ncol=8 columnlabels
ParamName StartValue GroupName BetaAssoc SenMethod x1 x2 x3
P1 2.000000000000000E-04 pargp_uno 1 1 1.0000E+00 1.000E+000.00E+00
P2 2.000000000000000E-04 pargp_uno 1 1 2.0000E+00 1.000E+000.00E+00
P3 2.000000000000000E-04 pargp_uno 1 1 3.0000E+00 1.000E+000.00E+00
...
P440 2.000000000000000E-04 pargp_uno 1 1 2.0000E+01 2.100E+010.00E+00
P441 2.000000000000000E-04 pargp_uno 1 1 2.1000E+01 2.100E+010.00E+00
END parameter_data
```

obs12.txt

```
BEGIN observation_data TABLE
nrow=12 ncol=4 columnlabels ObsName ObsValue GroupName Weight
P001T0000 33.8154 obsgp_oden 1.0
```

```
P002T0000 29.9383 obsgp_oden 1.0
P003T0000 28.4674 obsgp_oden 1.0
P004T0000 30.9286 obsgp_oden 1.0
P005T0000 24.7332 obsgp_oden 1.0
P006T0000 31.5769 obsgp_oden 1.0
P007T0000 27.3057 obsgp_oden 1.0
P008T0000 29.3834 obsgp_oden 1.0
P009T0000 27.8658 obsgp_oden 1.0
P010T0000 30.4177 obsgp_oden 1.0
P011T0000 28.5865 obsgp_oden 1.0
P012T0000 27.4403 obsgp_oden 1.0
END observation_data
```

Tables 2.1 and 2.2 summarize the input blocks and variables names, types, and default values.

Reference Cited

Doherty, J., 2010, PEST, Model-independent parameter estimation—User manual (5th ed., with slight additions): Brisbane, Australia, Watermark Numerical Computing.

Table 2.1.   Summary of input blocks with variables identified.

algorithmic cv	Algorithmic Control Variables: KEYWORDS		*This block is optional if all default values are used*
Variable type	*Variable name*	*Default*	*Description*
double precision	structural_conv	0.001	Structural parameter convergence values
double precision	phi_conv	0.001	Objective function convergence value
double precision	bga_conv	0.001	BGA outer loop convergence value
integer	it_max_structural	10	Max number of iterations for struct parameters
integer	it_max_phi	10	Max number of iterations for objective function
integer	it_max_bga	10	Max number of iterations for BGA
integer	linesearch	0	Linesearch procedure flag: [0] not perform [1] perform
integer	it_max_linesearch	10	Max number of iterations for linesearch procedure
integer	theta_cov_form	0	Form of theta covariance: [0] none, [1] diag, [2] full matrix
integer	deriv_mode	0	Derivatives (Jacobian) calculation method: [0] make PEST files internally, [1] use secondary command line argumern (typically adjoint state)
integer	posterior_cov_flag	0	[0] do not calculate posterior covariance, [1] calculate posterior covariance
character, len = 6	jacobian_format	"binary"	Two options for how the Jacobian matrix calculated by an external code is communicated to bgaPEST: binary means a jco file, ascii means a text file
character(len=100)	jacobian_file	scratch.jco	Jacobian File
integer	par_anisotropy	0	Anisotropy flag: [0] no anistropy, [1] anisotropy
integer	Q_compression_flag	0	[0] none - calculate full Qss, [1] Calculate Qss for each beta separately and if nugget store just 1, if toep_flag store just a vector
prior mean cv	Prior Mean Control Variables: KEYWORDS		
Variable type	*Variable name*	*Default*	*Description*
integer	prior_betas	0	Have or not prior informations about mean? [0] No - [1] Yes
integer	beta_cov_form	0	Form of Beta covariance: [0] none, [1] diag, [2] full matrix
prior mean data	Beta Association Data: TABLE		
Variable type	*Variable name*	*Defaults*	*Description*
integer	BetaAssoc	-	Integer identifiers of beta associations
character(len=100)	Partrans	-	Vector of parameter transformation : [NONE], [POWER], or[LOG]. (Not case sensitive)
double precision	beta_0	-	Prior beta values
double precision	beta_cov_i i = 1, p	-	Covariance of beta
structural parameter cv	Structural Parameter Control Variables: TABLE		
Variable type	*Variable name*	*Default*	*Description*
integer	BetaAssoc	-	Integer identifiers of beta associations
integer	prior_cov_mode	1	Supplied matrix [0] or calculated [1].
integer	var_type	1	Type of variogram [0] pure nugget, [1] linear, [2] exponential
integer	struct_par_opt	1	Structural parameters optimization: [0] No optimization, [1] Optimization
integer	trans_theta	1	Transformation of structural parameters in the estimation space (power transform): [0] No, [1] Yes
double precision	alpha_trans	50	Exponent of power transformation in case of trans_theta
structural parameter data	Structural Parameter Data: TABLE		
Variable type	*Variable name*	*Default*	*Description*
integer	BetaAssoc	-	Integer identifiers of beta associations
double precision	theta_0_1	-	Initial value of theta 1 value
double precision	theta_0_2	-	Initial value of theta 2 value -- negative if not used
structural parameter cov	Structural Parameter Data: TABLE		
Variable type	*Variable name*	*Default*	*Description*
integer	BetaAssoc	-	Integer identifiers of beta associations
double precision	theta_cov_i i=1,...,max(num_thata_type)	-	Theta covariance matrix
epistemic error term	Epistemic Error Term: KEYWORDS		
Variable type	*Variable name*	*Defaults*	*Description*
double precision	sig_0	-	Initial value of sigma (epistemic uncertainty parameter)
integer	sig_opt	-	Optimization for sigma: [0] No, [1] Yes
double precision	sig_p_var	0	Prior variance on sigma
integer	trans_sig	0	Transformation of epistemic error in the estimation space (power transform): [0] No, [1] Yes
double precision	alpha_trans_sig	50	Exponent of power transformation in case of trans_sig

Table 2.2. Summary of input blocks with variables identified (continued).

parameter cv	Parameter Control Variables: KEYWORDS		
Variable type	*Variable name*	*Defaults*	*Description*
integer	ndim	-	Spatial dimensions for parameters (1 if temporal only)
Q compression cv	Prior Covariance Compression Control Variables: TABLE		
Variable type	*Variable name*	*Defaults*	*Description*
integer	BetaAssoc	-	Integer identifiers of beta associations
integer	Toep_flag	-	Using Toeplitz matrix for Qss. [0] No, [1] Yes
integer	Nrow	-	Number of model rows
integer	Ncol	-	Number of model columns
integer	Nlay	-	Number of model layers
parameter groups	Parameter Groups: TABLE		
Variable type	*Variable name*	*Defaults*	*Description*
character (len=50)	groupname	-	Name of the parameter groups
integer	grouptype	-	Identifier to segregate groups of different types
double precision	derinc	-	Derivative increment for external Jacobian
parameter data	Parameter Data: TABLE		
Variable type	*Variable name*	*Defaults*	*Description*
character (len=50)	GroupName	-	Name of group
double precision	StartValue	-	Starting values of parameters
character (len=50)	ParamName	-	Name of parameter
double precision	x1	-	Location in first dimension (time if a time series)
double precision	x2	-	Location in second dimension (read if ndim >= 2)
double precision	x3	-	Location in third dimension (read if ndim >= 3)
integer	SenMethod	-	Sensitivity calculation method
integer	BetaAssoc	-	Beta association
observation groups	Observation Groups: TABLE		
Variable type	*Variable name*	*Defaults*	*Description*
character (len=50)	groupname	-	Name of the observation groups
observation data	Observation Data: TABLE		
Variable type	*Variable name*	*Defaults*	*Description*
character (len=50)	GroupName	-	Name of groups
double precision	ObsValue	-	Vector of observations
character (len=50)	ObsName	-	names of observations
double precision	Weight	-	Weight for R matrix
model command lines	Model Command Lines: KEYWORDS		
Variable type	*Variable name*	*Defaults*	*Description*
character (len=50)	Command	-	Command line
character (len=50)	DerivCommand	-	Derivative Command line
model input files	Model Input Files: TABLE		
Variable type	*Variable name*	*Defaults*	*Description*
character(len=100)	TemplateFile	-	Template file
character(len=100)	ModInFile	-	Input file
model output files	Model Input Files: TABLE		
Variable type	*Variable name*	*Defaults*	*Description*
character(len=100)	InstructionFile	-	Instruction file
character(len=100)	ModOutFile	-	Output file
parameter anisotropy	Parameter Anisotropy: TABLE	*This block is optional if parameter anisotropy is not used*	
Variable type	*Variable name*	*Defaults*	*Description*
integer	BetaAssoc	-	Integer identifiers of beta associations
double precision	horiz_angle	-	Angle, in degrees, of principal anisotropy direction
double precision	horiz_ratio	-	Ratio of maximum to minimum principal property values in the horizontal plane
double precision	vertical_ratio	-	Ratio of maximum to minimum principal property values in the vertical direction (read only if ndim=3)

Appendix 3—Details of the Method

The Bayesian geostatistical approach is described in detail by Kitanidis and Vomvoris (1983), Hoeksema and Kitanidis (1984), Kitanidis (1995), and Nowak and Cirpka (2004) among others. The mathematics are reviewed here.

The Bayesian Geostatistical Approach

In the Bayesian geostatistical approach, the posterior pdf is calculated as

$$p(\mathbf{s}|\mathbf{y}) \propto \underbrace{\exp\left(-\frac{1}{2}(\mathbf{y}-\mathbf{h}(\mathbf{s}))^T \mathbf{R}^{-1}(\mathbf{y}-\mathbf{h}(\mathbf{s}))\right)}_{L(\mathbf{y}|\mathbf{s})} \underbrace{\exp\left(-\frac{1}{2}(\mathbf{s}-\mathbf{X}\beta^*)^T \mathbf{G}_{ss}^{-1}(\mathbf{s}-\mathbf{X}\beta^*)\right)}_{p(\mathbf{s})} \tag{3.1}$$

where $\mathbf{s}$ is the $m \times 1$ vector of parameter values at distributed spatial locations in the model, $\mathbf{X}\beta^*$ is the prior mean, $\mathbf{G}_{ss}$ is the $m \times m$ prior covariance of $(\mathbf{s}-\mathbf{X}\beta^*)$, $\mathbf{h}(\mathbf{s})$ is the $n \times 1$ vector of modeled forecasts colocated with observations $(\mathbf{y})$, and $\mathbf{R}$ is the $n \times n$ epistemic uncertainty covariance, modeled as $\sigma_R^2 \mathbf{W}$ where σ_R^2 represents epistemic uncertainty, and $\mathbf{W}$ is an $n \times n$ diagonal matrix containing the observation weights. In general terms, the likelihood function $(L(\mathbf{y}|\mathbf{s}))$ characterizes the misfit between model forecasts and observations whereas, the prior pdf $(p(\mathbf{s}))$ defines a characteristic (such as smoothness or continuity) that is assumed to apply to the parameter field. The prior pdf also serves the role of regularization. The best estimate of $\mathbf{s}$ maximizes the posterior pdf. A computationally efficient method to find the best estimates of $\mathbf{s}$ and β $\left(\hat{\mathbf{s}}\text{ and }\hat{\beta}\text{, respectively}\right)$ is through

$$\hat{\mathbf{s}} = \mathbf{X}\hat{\beta} + \mathbf{Q}_{ss}\mathbf{H}^{\mathbf{T}}\xi \tag{3.2}$$

which is the superposition of the prior mean (first term) and an innovation term that accounts for deviations of the model outputs from the observations (second term). In this context, β in the first term is not the prior mean but is the best estimate of the mean (mapped onto the parameter field through the $m \times p$ $\mathbf{X}$ matrix), whereas the second term is fluctuations about the estimated mean. $\mathbf{H}$ in the second term (often referred to as the $n \times m$ Jacobian, sensitivity, or susceptibility matrix) is the sensitivity of observation values to parameter values where $H_{ij} = \frac{\partial \mathbf{h}(\mathbf{s})_i}{\partial \mathbf{s}_j}$, which can be calculated by using either finite difference or adjoint-state methods. In bgaPEST, finite-difference calculations for $\mathbf{H}$ are made by using PEST, whereas adjoint-state calculations depend on the specific model being used and must be made by using an external program.

The values for $\hat{\beta}$ and ξ are found by solving the $(n+p) \times (n+p)$ linear system of cokriging equations

$$\begin{bmatrix} \mathbf{Q}_{yy} & \mathbf{HX} \\ \mathbf{X}^T\mathbf{H}^T & -\mathbf{Q}_{\beta\beta}^{-1} \end{bmatrix} \begin{bmatrix} \xi \\ \hat{\beta} \end{bmatrix} = \begin{bmatrix} \mathbf{y} \\ -\mathbf{Q}_{\beta\beta}^{-1}\beta^* \end{bmatrix} \tag{3.3}$$

where $\mathbf{Q}_{yy}$ is the $n \times n$ auto-covariance matrix of the observations, defined as $\mathbf{HQ}_{ss}\mathbf{H}^{\mathbf{T}} + \mathbf{R}$, n is the number of observations, and p is the number of beta associations. In hydrogeologic applications, the numerical forward model is typically nonlinear. Further nonlinearity can be induced by using a logarithmic or power transformation, which is a convenient way to enforce non-negativity on parameters.

Provided that the nonlinearities introduced are not too extreme, a solution can be obtained through successive linearizations following the quasi-linear extension (Kitanidis, 1995). The forward model, $\mathbf{h}(\mathbf{s})$ is expanded about the current best estimate of the parameters $\tilde{\mathbf{s}}$

$$\mathbf{h}(\mathbf{s}) \approx \mathbf{h}(\tilde{\mathbf{s}}) + \tilde{\mathbf{H}}(\mathbf{s}-\tilde{\mathbf{s}}) \tag{3.4}$$

where $\tilde{\mathbf{H}}$, as a function of $\tilde{\mathbf{s}}$, is evaluated at each linearization. We assign the subscript $_k$ to indicate iteration number, and correct the measurements for the kth linearization as

$$\mathbf{y}_k' = \mathbf{y} - \mathbf{h}(\tilde{\mathbf{s}}_k) + \tilde{\mathbf{H}}_k \tilde{\mathbf{s}}_k. \tag{3.5}$$

Then the cokriging equations (equation 3.3) are updated

$$\begin{bmatrix} \tilde{\mathbf{Q}}_{\mathbf{yy},k} & \tilde{\mathbf{H}}_k \mathbf{X} \\ \mathbf{X}^T \tilde{\mathbf{H}}_k^T & -\mathbf{Q}_{\beta\beta}^{-1} \end{bmatrix} \begin{bmatrix} \xi_k \\ \hat{\beta}_k \end{bmatrix} = \begin{bmatrix} \mathbf{y}_k' \\ -\mathbf{Q}_{\beta\beta}^{-1} \beta^* \end{bmatrix} \tag{3.6}$$

where $\tilde{\mathbf{Q}}_{\mathbf{yy},k} = \tilde{\mathbf{H}}_k \mathbf{Q}_{\mathbf{ss}} \tilde{\mathbf{H}}_k^T + \mathbf{R}$. From this set of equations, the next estimate of $\mathbf{s}$ is

$$\tilde{\mathbf{s}}_{k+1} = \mathbf{X}\hat{\beta}_k + \mathbf{Q}_{\mathbf{ss}} \tilde{\mathbf{H}}_k^T \xi_k. \tag{3.7}$$

This process can be iterated until there is minimal difference in the parameter estimates or minimal further improvement in the objective function. The objective function, which we seek to minimize, is $-\ln p(\mathbf{s}|\mathbf{y})$ which is equivalent to maximizing equation 3.1

$$\Phi_T = -\frac{1}{2} \ln p(\mathbf{s}|\mathbf{y}) = (\mathbf{s} - \mathbf{X}\beta^*)^T \mathbf{G}_{\mathbf{ss}}^{-1} (\mathbf{s} - \mathbf{X}\beta^*) - \frac{1}{2} (\mathbf{y} - \mathbf{h}(\mathbf{s}))^T \mathbf{R}^{-1} (\mathbf{y} - \mathbf{h}(\mathbf{s})) \tag{3.8}$$

Line Search

In some cases, numerical instability makes convergence difficult. A line search is implemented in which a linear search is performed between the most recent best estimate of the parameters ($\hat{\mathbf{s}}$) and the current linearization of the parameters ($\tilde{\mathbf{s}}$), seeking a parameter value that minimizes an objective function.

The line search optimizes a single parameter, ρ, along a linear dimension between $\hat{\mathbf{s}}$ and $\tilde{\mathbf{s}}$ as

$$\mathbf{s}_{opt} = \hat{\mathbf{s}}\rho + \tilde{\mathbf{s}}(1 - \rho) \tag{3.9}$$

where $\mathbf{s}_{opt}$ minimizes the objective function, Φ_T, using a Nelder-Mead simplex (see, for example, Press and others, 1992), which guarantees monotonic decrease in Φ_T over successive iterations. It is recommended to limit the number of line-search iterations to a relatively low number, because the goal of handling weak linearity is balanced against the computations required to perform the line search. The greatest advantage is likely achieved in the first few (less than five) iterations. The role of the line search is not to find a minimum value of Φ_T because the nonlinearity of the overall problem prevents it. Rather, the line search is meant to be a correction of search direction for stability.

Parameter Field Anisotropy

In distributed parameter fields, such as hydraulic conductivity in groundwater models, it is common to encounter anisotropy along an axis that may or not be aligned with the coordinate (x, y, z) axes. bgaPEST allows the definition of anisotropy in a horizontal plane at any angle from the x-axis and also in the vertical direction. The general layout of horizontal anisotropy is illustrated in figure 3.1. The angle from the x-axis (specified in degrees) is defined by `horiz_angle`, and the amount of anisotropy is defined by `horiz_ratio`. The designation `p_max` refers to the direction with maximum parameter values and `p_min` refers to the direction of minimum parameter values. The ratio is used to adjust the effective distance (and thereby the covariance values) along that principal direction. The user supplies values for `horiz_angle` and `horiz_ratio` for each beta association. If some beta associations are not meant to exhibit anisotropy, the user may simply set `horiz_ratio=1.0`. If none of the beta associations exhibit anisotropy, the entire `parameter_anisotropy`

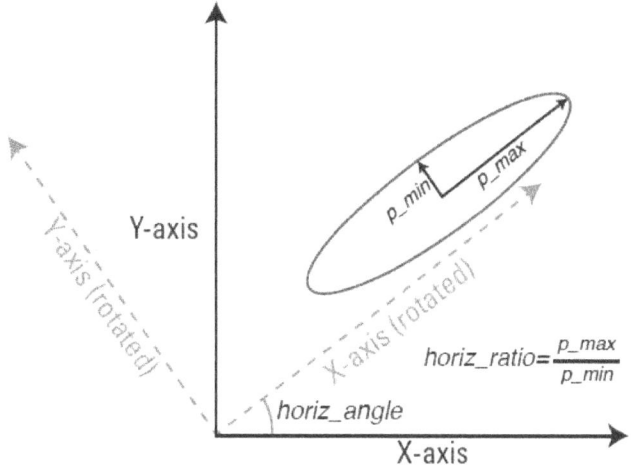

Figure 3.1. Schematic diagram of conventions defining horizontal anisotropy in bgaPEST .

block can be eliminated by setting the algorithmic control variable par_anisotropy=0, which means the block (if present) is ignored.

Anisotropy is introduced in the calculation of distances that are then used in the calculation of the prior covariance matrix $\mathbf{Q}_{ss}$ discussed below. Every pair of points must first be rotated into the principal direction orientation. This is accomplished by means of a rotation matrix:

$$\begin{bmatrix} x_{rot,i} \\ y_{rot,i} \end{bmatrix} = \begin{bmatrix} \cos\theta & -\sin\theta \\ \sin\theta & \cos\theta \end{bmatrix} \begin{bmatrix} x_i \\ y_i \end{bmatrix} \tag{3.10}$$

where i indicates the ith point of the pair ($i = 1, 2$), θ is the angle (in degrees) specified by horiz_angle, x and y constitute the point coordinates in the original coordinate system, and x_{rot} and y_{rot} constitute the location projected into the coordinate system corresponding to the orientation of horizontal anisotropy. Once this projection is made, the horizontal distance is calculated as

$$distance = \sqrt{\left(x_{rot,1} - x_{rot,2}\right)^2 + \text{horiz\_ratio} \times \left(y_{rot,1} - y_{rot,2}\right)^2} \tag{3.11}$$

For three-dimensional parameter fields, a second anisotropy ratio may be specified as vertical_ratio. In the vertical direction, no angle is specified, so the rotation step is not required and distance is calculated as

$$distance = \sqrt{\begin{array}{l} \left(x_{rot,1} - x_{rot,2}\right)^2 \\ + \text{horiz\_ratio} \times \left(y_{rot,1} - y_{rot,2}\right)^2 \\ + \text{vertical\_ratio} \times \left(z_{rot,1} - z_{rot,2}\right)^2 \end{array}} \tag{3.12}$$

Prior Probability Density Function

The prior pdf of s can be characterized as multi-Gaussian through its mean and covariance. The $(m \times 1)$ unknown parameter vector, **s**, is modeled as a random process with mean

$$E[\mathbf{s}] = \mathbf{X}\beta \tag{3.13}$$

where $E[\cdot]$ indicates expected value, m is the number of parameters, β is a $(p \times 1)$ vector of drift coefficients, and $\mathbf{X}$ is an $(m \times p)$ matrix of base functions. In the absence of prior drift, the β constitute the beta association mean values, and $\mathbf{X}$ is a selection matrix mapping each value in $\mathbf{s}$ and β into their appropriate beta association. $\mathbf{X}$ contains all zeros except for $\mathbf{X}_{ij}^{th}$ element, which maps the ith parameter to the jth beta association, which contains the value of 1. Subdivision into beta associations within distributed parameter domains has been critical to success in hydrogeologic settings that include strong contrasts in parameter values indicative of geologic contacts (see, for example, Fienen and others, 2004; Fienen and others, 2008). Prior drift is accounted for in $\mathbf{X}$ through trends expressed in the nonzero terms, although this is not currently implemented in bgaPEST. The prior covariance $(\mathbf{Q}_{ss})$ of $\mathbf{s}$ for a known β is

$$\mathbf{Q}_{ss}(\theta) = E[(\mathbf{s}-\mathbf{X}\beta)(\mathbf{s}-\mathbf{X}\beta)^T] \tag{3.14}$$

where $\mathbf{Q}_{ss}$ is a covariance function with structural parameters θ. In bgaPEST, allowable covariance functions include

1. the nugget

$$\mathbf{Q}_{ss}(\mathbf{d}) = \sigma^2 \tag{3.15}$$

and

2. the exponential covariance function

$$\mathbf{R}_{ss}(\mathbf{d}) = \sigma^2 \exp(-\frac{|\mathbf{d}|}{\ell}) \tag{3.16}$$

where $|\mathbf{d}|$ is separation distance, σ^2 is variance, and ℓ is integral scale. If the integral scale is set such that $\ell > \max(|\mathbf{d}|)$ we can substitute $\sigma^2 = \theta\ell$ and restate equation 3.16 as

$$\mathbf{Q}_{ss}(\mathbf{h}, \sigma^2) = \sigma^2 \ell \exp(-\frac{|\mathbf{d}|}{\ell}). \tag{3.17}$$

We can also set $\ell = 10 \times \max(|\mathbf{d}|)$ so that the behavior of the covariance function will be as a linear variogram (Fienen and others, 2008) which enforces continuity at a scale determined by the single free structural parameter σ^2. The motivation for this covariance function choice is to impart minimal assumptions about parameter structure onto the solution.

The appropriate values of θ (the vector of structural parameters including σ_R^2 and σ^2) are calculated through restricted maximum likelihood. For the remainder of this derivation, θ is assumed to be known. In bgaPEST, as discussed below in the input instructions, either the exponential or linear variogram models may be used.

Assembling the mean and covariance, the prior pdf is

$$p(\mathbf{s}|\beta) \propto \exp\left[-\frac{1}{2}(\mathbf{s}-\mathbf{X}\beta)^{\mathrm{T}}\mathbf{Q}_{ss}^{-1}(\mathbf{s}-\mathbf{X}\beta)\right]. \tag{3.18}$$

In the case of no knowledge about the prior mean the prior pdf of β can be modeled as uniform over all space as $p(\beta) \propto \mathbf{1}$ with both $\mathbf{s}$ and β being estimated together, so that the conditional distribution in equation 3.18 is replaced by a joint distribution

$$p(\mathbf{s}, \beta) \propto \exp\left[-\frac{1}{2}(\mathbf{s}-\mathbf{X}\beta)^{\mathrm{T}}\mathbf{Q}_{ss}^{-1}(\mathbf{s}-\mathbf{X}\beta)\right]. \tag{3.19}$$

Frequently, at least diffuse knowledge about the prior mean is available and can be modeled as multi-Gaussian with mean β^* and covariance $\mathbf{Q}_{\beta\beta}$. Typically, $\mathbf{Q}_{\beta\beta}$ is modeled as a diagonal matrix with variance values on the diagonal indicating independence among the β^*. Incorporating the prior information yields a prior pdf for $\mathbf{s}$

$$p(\mathbf{s}) \propto \exp\left[-\frac{1}{2}(\mathbf{s}-\mathbf{X}\beta^*)^{\mathrm{T}}\mathbf{G}_{\mathbf{ss}}^{-1}(\mathbf{s}-\mathbf{X}\beta^*)\right] \tag{3.20}$$

where $\mathbf{X}\beta^*$ is the prior mean and $\mathbf{G}_{\mathbf{ss}} = \mathbf{Q}_{\mathbf{ss}} + \mathbf{X}\mathbf{Q}_{\beta\beta}\mathbf{X}^{\mathrm{T}}$ is the prior covariance (Nowak and Cirpka, 2004). The incorporation of prior mean information, even assuming very high variance values in $\mathbf{Q}_{\beta\beta}$, can provide numerical stability without overly biasing the results. In the original formulations of Kitanidis and Vomvoris (1983), Hoeksema and Kitanidis (1984), and Kitanidis (1995), $\mathbf{G}_{\mathbf{ss}} = \mathbf{Q}_{\mathbf{ss}}$ and no prior covariance is supplied on the values for β. This behavior can be duplicated in bgaPEST by specifying prior_betas=0 in the input block for Prior Mean Control Variables.

Prior Covariance Matrix Storage Issues

In underdetermined problems suitable for bgaPEST, the number of parameters can be very large. The prior covariance matrix discussed above can, therefore, grow to such large dimensions that it cannot be practically stored in computer memory. However, two techniques are provided to alleviate some of this storage stress: compression and Toeplitz transformation.

Compression takes advantage of the fact that values in the $\mathbf{G}_{\mathbf{ss}}$ matrix relating parameters in different beta associations, by definition, have the value of zero. As a result, a general $\mathbf{G}_{\mathbf{ss}}$ matrix can be viewed as a partitioned matrix of nonzero blocks $\left(\mathbf{G}_{\mathbf{ss},\beta i}\right)$ and zero blocks

$$\mathbf{G}_{\mathbf{ss}} = \begin{bmatrix} \mathbf{G}_{\mathbf{ss},\beta 1} & \mathbf{0} & \mathbf{0} & \mathbf{0} \\ \mathbf{0} & \mathbf{G}_{\mathbf{ss},\beta 2} & \mathbf{0} & \mathbf{0} \\ \mathbf{0} & \mathbf{0} & \ddots & \mathbf{0} \\ \mathbf{0} & \mathbf{0} & \mathbf{0} & \mathbf{G}_{\mathbf{ss},\beta p} \end{bmatrix}. \tag{3.21}$$

There is no need to store the zero elements provided that accommodations are made to avoid multiplications that involve the zeros. These accommodations have been made in bgaPEST and compression is, therefore, allowed.

In cases where spacing of model cells or nodes is constant in respective directions, $\mathbf{G}_{\mathbf{ss},\beta i}$ is a block Toeplitz matrix (Gray, 2005). A square, symmetric $j \times j$ matrix is Toeplitz in form if it has diagonals that all have the same value as in this example

$$\mathbf{T} = \begin{bmatrix} t_0 & t_1 & \cdots & t_{j-2} & t_{j-1} \\ t_1 & t_0 & t_1 & \ddots & t_{j-2} \\ \vdots & t_1 & t_0 & \ddots & \vdots \\ t_{j-2} & \ddots & \ddots & \ddots & t_1 \\ t_{j-1} & t_{j-2} & \cdots & t_1 & t_0 \end{bmatrix}. \tag{3.22}$$

This matrix has the properties that there are only j unique values, and these values occur in a regular order such that only a vector of length j needs to be stored from which individual rows can be constructed to perform matrix multiplication operations. The spacing, as indicated above, must be constant. For example, in a spatial model such as properties in a groundwater model, Δx, Δy, and Δz must be constant, but these values do not need to be equal to each other. This condition is restrictive in the sense that it implies a regular grid that may

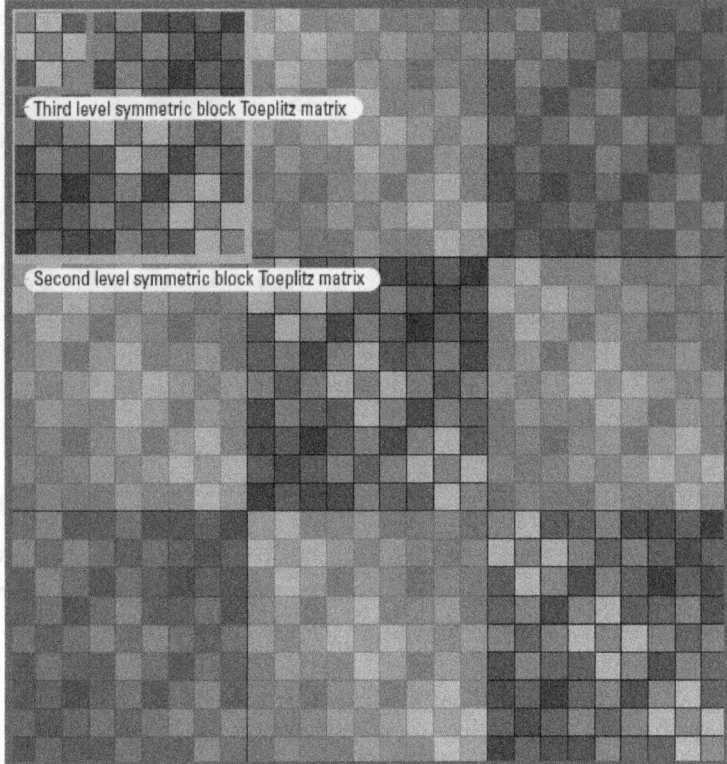

Figure 3.2.   Schematic graphical visualization of a three-level embedded set of Toeplitz blocks in a covariance matrix (modified from D'Oria (2010)). The smallest squares represent individual matrix entries (values), and the colors correspond to distinct values. In this synthetic example, it is assumed that there are three rows, three columns, and three layers in the underlying model.

not correspond to geometry in the field; however, the regular grid required to take advantage of Toeplitz storage and operations can be assigned to one beta association with a surrounding, irregular grid put in another beta association with fewer parameter values.

In order to use Toeplitz structure in three dimensions, there must be a three-level embedding of Toeplitz matrices (D'Oria, 2010). The first level corresponds to the model layers, the second level corresponds to model rows, and the third level corresponds to model columns. Inspection of the schematic example in figure 3.2 shows that every distinct value represented in the entire matrix is found in the first (leftmost) column. Cycling of rows or columns, relative to the single stored vector, can be used to reconstruct any row or column of the main matrix to be used in multiplication operations. In bgaPEST, a combination of Toeplitz and complete blocks can make up the $\mathbf{G_{ss}}$ matrix in Compressed form, as discussed above.

For even larger problems (more parameters) specialized Fourier transform-based functions may be useful to speed up the computations made with compressed matrices (Nowak and others, 2003; Nowak and Cirpka, 2004). The restriction of regular grid spacing can also be relaxed by using a Karhunen-Loève transform (Li and Cirpka, 2006). Both of these advances may be considered as future improvements to bgaPEST but are not currently implemented.

Detailed input instructions for bgaPEST are presented in appendix 1. It is important to note, however, that if Toeplitz compression is invoked, parameters must be listed in the `.bgp` input file in order, sorted first by layer, then by column, and finally by row.

Likelihood Function

The parameters **s** are related to observations **y** through a measurement equation

$$\mathbf{y} = \mathbf{h}(\mathbf{s}) + \mathbf{v} \tag{3.23}$$

where **y** is an $(n \times 1)$ vector of observations, such as hydraulic heads or solute concentrations, $\mathbf{h}(\mathbf{s})$ is a transfer function or numerical model that calculates predictions which are colocated spatially and temporally with the observation values, and **v** is an $(n \times 1)$ vector of epistemic uncertainty terms, modeled as a random process with zero mean and covariance matrix **R**. Epistemic uncertainty is the result of imperfect or sparse measurements and an incomplete or inappropriate conceptual model (Rubin, 2003, p. 4). The epistemic uncertainty terms are assumed to be independent and uncorrelated, so

$$\mathbf{R} = \sigma_{\mathbf{R}}^2 \mathbf{W} \tag{3.24}$$

where σ_R^2 is the epistemic uncertainty parameter and **W** is an $(n \times n)$ diagonal weight matrix in which each element is $\mathbf{W}_{ii} = \frac{1}{\omega_i^2}$ where ω_i is the ith weight, specified by the user. The purpose of the values of ω is to allow for different confidence in different individual observations or groups of observations. In reality, the component of epistemic uncertainty due to measurement error is likely uncorrelated, but the component due to modeling and conceptual uncertainty is likely systematic and correlated (Gaganis and Smith, 2001). A significant portion of this uncertainty may be reduced by not lumping parameters into homogeneous zones (Gallagher and Doherty, 2007), and the means to characterize the structure of **R** are rarely available. If information about **R** is available, however, it could be included and equation 3.24 replaced by a more complicated matrix. This option is currently not available in bgaPEST, however. Proceeding with equation 3.24, the likelihood function, assumed to be multi-Gaussian, is

$$L(\mathbf{y}|\mathbf{s}) \propto \exp\left[-\frac{1}{2}(\mathbf{y} - \mathbf{h}(\mathbf{s}))^{\mathrm{T}} \mathbf{R}^{-1}(\mathbf{y} - \mathbf{h}(\mathbf{s})) \right]. \tag{3.25}$$

The structural parameter for the likelihood function is σ_R^2 and is calculated along with θ by using restricted maximum likelihood.

Posterior Probability Density Function

Applying Bayes' theorem with the product of equations 3.20 and 3.25 yields the posterior pdf

$$p(\mathbf{s}|\mathbf{y}) \propto \exp\left[-\frac{1}{2}(\mathbf{s} - \mathbf{X}\beta^*)^{\mathrm{T}} \mathbf{G}_{\mathbf{ss}}^{-1}(\mathbf{s} - \mathbf{X}\beta^*) - \frac{1}{2}(\mathbf{y} - \mathbf{h}(\mathbf{s}))^{\mathrm{T}} \mathbf{R}^{-1}(\mathbf{y} - \mathbf{h}(\mathbf{s})) \right] \tag{3.26}$$

The best estimate of **s** maximizes the posterior pdf. A computationally efficient method to find the best estimates of **s** and β $\left(\hat{\mathbf{s}} \text{ and } \hat{\beta}, \text{respectively} \right)$ is through

$$\hat{\mathbf{s}} = \mathbf{X}\hat{\beta} + \mathbf{Q}_{\mathbf{ss}}\mathbf{H}^{\mathrm{T}}\xi \tag{3.27}$$

which is the superposition of the prior mean (first term) and an innovation term that accounts for deviations of the model outputs from the observations (second term). **H** in the second term (often referred to as the Jacobian, sensitivity, or susceptibility matrix) is the sensitivity of observation values to parameter values where $H_{ij} = \frac{\partial \mathbf{h}(\mathbf{s})_i}{\partial \mathbf{s}_j}$ is calculated by using either finite-difference or adjoint-state methods. The values for $\hat{\beta}$ and

ξ are found by solving the $(n+p) \times (n+p)$ linear system of cokriging equations

$$
\begin{bmatrix} \mathbf{Q}_{yy} & \mathbf{HX} \\ \mathbf{X}^T \mathbf{H}^T & -\mathbf{Q}_{\beta\beta}^{-1} \end{bmatrix} \begin{bmatrix} \xi \\ \hat{\beta} \end{bmatrix} = \begin{bmatrix} \mathbf{y} \\ -\mathbf{Q}_{\beta\beta}^{-1}\beta^* \end{bmatrix} \tag{3.28}
$$

where $\mathbf{Q}_{yy}$ is the auto-covariance matrix of the observations, defined as $\mathbf{HQ}_{ss}\mathbf{H}^T + \mathbf{R}$.

Quasi-Linear Extension

As discussed by Kitanidis (1995), we must adjust calculations of the posterior pdf to account for nonlinearity. To do this, we expand the solution in a first-order Taylor expansion, resulting in an updated set of cokriging equations from equation 3.28:

$$
\begin{bmatrix} \mathbf{Q}_{yy} & \mathbf{HX} \\ \mathbf{X}^T \mathbf{H}^T & -\mathbf{Q}_{\beta\beta}^{-1} \end{bmatrix} \begin{bmatrix} \xi \\ \hat{\beta} \end{bmatrix} = \begin{bmatrix} \mathbf{y} - \mathbf{h}(\tilde{\mathbf{s}}) + \mathbf{H}\tilde{\mathbf{s}} \\ -\mathbf{Q}_{\beta\beta}^{-1}\beta^* \end{bmatrix} \tag{3.29}
$$

At each iteration (later referred to as inner iterations), the system in equation 3.29 is solved, resulting in an updated estimate of $\hat{\mathbf{s}}$ calculated through equation 3.27. At each iteration, the objective function, based on minimizing the negative logarithm of the posterior pdf (equation 3.26) is evaluated by using the current value of $\hat{\mathbf{s}}$: this is equivalent to finding the values of $\mathbf{s}$ that *maximize* the posterior probability. Switching to a minimization problem and taking the logarithm has computational advantages.

The objective function, then, is

$$
\Phi_T = \Phi_M + \Phi_R \tag{3.30}
$$

where Φ_T is the total objective function, Φ_M is the misfit objective function (also corresponding to the likelihood function) and Φ_R is the regularization objective function (also corresponding to the prior pdf). The components of equation 3.30 are

$$
\Phi_M = \frac{1}{2}(\mathbf{y} - \mathbf{h}(\hat{\mathbf{s}}))^T \mathbf{R}^{-1}(\mathbf{y} - \mathbf{h}(\hat{\mathbf{s}})) \tag{3.31}
$$

and

$$
\Phi_R = \frac{1}{2}(\hat{\mathbf{s}} - \mathbf{X}\beta^*)^T \mathbf{G}_{ss}^{-1}(\hat{\mathbf{s}} - \mathbf{X}\beta^*) \tag{3.32}
$$

where both the negative signs and exponentiation are obviated by taking the negative logarithm of $p(\mathbf{s}|\mathbf{y})$.

Implementation of Partitions into Beta Associations

The concept of beta associations is discussed above, and details of their implementation are given here. First, the prior covariance matrix $\mathbf{Q}_{ss}$ is censored by assigning a value of zero to each element that characterizes covariance between cells of different regions or parameter types, as defined by beta associations. It is not required that the covariance model be the same for each beta association. If different covariance models are used for different zones, this is reflected in the appropriate parts of $\mathbf{Q}_{ss}$. Furthermore, in some applications, a single structural parameter, θ, may be estimated and applied to all of $\mathbf{Q}_{ss}$. In other cases, and necessarily if the covariance model differs in various beta association, multiple elements of θ are estimated.

A distinct prior mean parameter β^* is assigned for each beta association, and the matrix $\mathbf{X}$ (equation 3.13) is determined as explained above. In cases where the mean of each zone is completely unknown, no values for

β^* are provided, but the **X** matrix is constructed nonetheless and in both cases a value of $\hat{\beta}$ is calculated for each beta association.

Structural Parameters and Restricted Maximum Likelihood

A vital element to the method outlined above is proper selection of the structural parameters. Structural parameters—also called hyperparameters or nuisance parameters—are the parameters that characterize the covariance structure of both the epistemic uncertainty related to the observations, and the inherent variability of the parameters. In bgaPEST, structural parameters may include the epistemic uncertainty term in equation 3.24 $\left(\sigma_R^2\right)$ and the prior pdf variogram parameters in equation 3.17 (θ). These parameters are estimated by using restricted maximum likelihood consistent with the approaches of Kitanidis and Vomvoris (1983), Kitanidis (1995) and Li and others (2007).

Applying Bayes' theorem to the structural parameters, given the measurements, we calculate

$$p\left(\theta|\mathbf{y_k'}\right) \propto L\left(\mathbf{y}_k'|\theta\right) p\left(\theta\right) \tag{3.33}$$

The likelihood function evaluates how closely the observations and predictions match, given the current linearization and the current set of structural parameters

$$L\left(\mathbf{y}_k'|\theta\right) \propto \det\left(\mathbf{G_{yy}}\right)^{-\frac{1}{2}} \exp\left[-\frac{1}{2}\left(\mathbf{y}_k' - \mathbf{HX}\beta^*\right)^T \mathbf{G_{yy}^{-1}}\left(\mathbf{y}_k' - \mathbf{HX}\beta^*\right)\right] \tag{3.34}$$

where $\mathbf{G_{yy}}$ is the measurement auto-covariance defined as

$$\mathbf{G_{yy}} = \mathbf{Q_{yy}} + \mathbf{HXQ}_{\beta\beta}\mathbf{X}^T\mathbf{H}^T. \tag{3.35}$$

Note that $\mathbf{Q_{yy}}$ is intrinsically dependent upon the values of θ.

Prior information about the structural parameters may also be included, with prior mean θ^* and covariance matrix $\mathbf{Q}_{\theta\theta}$:

$$p\left(\theta\right) \propto \det\left(\mathbf{Q}_{\theta\theta}\right)^{-\frac{1}{2}} \exp\left[-\frac{1}{2}\left(\theta - \theta^*\right)^T \mathbf{Q}_{\theta\theta}^{-1}\left(\theta - \theta^*\right)\right] \tag{3.36}$$

The posterior pdf is the product of equations 3.36 and 3.34

$$p\left(\theta|\mathbf{y_k'}\right) \propto \det\left(\mathbf{Q}_{\theta\theta}\right)^{-\frac{1}{2}} \det\left(\mathbf{G_{yy}}\right)^{-\frac{1}{2}} \exp\left[\begin{array}{c} -\frac{1}{2}\left(\theta - \theta^*\right)^T \mathbf{Q}_{\theta\theta}^{-1}\left(\theta - \theta^*\right) \\ -\frac{1}{2}\left(\mathbf{y}_k' - \mathbf{HX}\beta^*\right)^T \mathbf{G_{yy}^{-1}}\left(\mathbf{y}_k' - \mathbf{HX}\beta^*\right) \end{array}\right]. \tag{3.37}$$

To find the most likely values for θ we minimize $-\ln\left(p\left(\theta|\mathbf{y_k'}\right)\right)$ resulting in the objective function

$$\Phi_S = \frac{1}{2}\ln\left(\det\left(\mathbf{G_{yy}}\right)\right) + \frac{1}{2}\left[\left(\theta - \theta^*\right)^T \mathbf{Q}_{\theta\theta}^{-1}\left(\theta - \theta^*\right) + \left(\mathbf{y}_k' - \mathbf{HX}\beta^*\right)^T \mathbf{G_{yy}^{-1}}\left(\mathbf{y}_k' - \mathbf{HX}\beta^*\right)\right] \tag{3.38}$$

where unchanging quantities are absorbed into the constant of proportionality including $\det\left(\mathbf{Q}_{\theta\theta}\right)^{-\frac{1}{2}}$. The optimal values for θ are found by using the Nelder-Mead simplex algorithm (for example, Press and others, 1992, p. 408–410). Non-negativity in the θ parameters can be enforced by using a power transformation (Box and Cox, 1964) discussed below. As indicated by Kitanidis (1995), nonlinearity requires that structural parameters to be estimated iteratively with the estimation of model parameters. This is accomplished through a sequence of coupled inversion as follows.

1. Initialize model parameters as $(\mathbf{s_0})$ and structural parameters (θ_0).

2. Solve for a new estimate of model parameters $(\hat{\mathbf{s}})$ holding θ constant.

3. Solve for a new estimate of structural parameters $\left(\hat{\theta}\right)$ holding $\mathbf{s}$ constant.

4. Repeat steps 2 and 3 until the change in θ in two consecutive outer iterations of steps 2 and 3 decreases below a specified tolerance.

Logarithmic and Power Transformations

In some cases, structural parameters and model parameters are best estimated in transformed space. A common reason is to enforce non-negativity. For model parameters either a logarithmic (base e) or a power transformation may be used. For structural parameters a power transformation is the only option.

The power transformation (Box and Cox, 1964; Fienen and others, 2004) is defined as:

$$\mathbf{s} = \alpha \left(\mathbf{p}^{\frac{1}{\alpha}} - 1 \right) \tag{3.39}$$

where $\mathbf{s}$ is the vector of transformed parameters, $\mathbf{p}$ is the vector of non-transformed parameters, and α is a tuning variable that controls the strength of the transformation. The back-transformation is:

$$\mathbf{p} = \left(\frac{\mathbf{s} + \alpha}{\alpha} \right)^{\alpha} \tag{3.40}$$

At the limit, as α increases to infinity, the transformation and back-transformation converge on the natural logarithm and exponential function, respectively:

$$\mathbf{s} = \lim_{\alpha \to \infty} = \alpha \left(\mathbf{p}^{\frac{1}{\alpha}} - 1 \right) = \ln(\mathbf{p}) \tag{3.41}$$

and

$$\mathbf{p} = \lim_{\alpha \to \infty} = \left(\frac{\mathbf{s} + \alpha}{\alpha} \right)^{\alpha} = \exp(\mathbf{s}). \tag{3.42}$$

Posterior Covariance

Calculation of the posterior covariance can be based on the inverse of the Hessian of the objective function (for example, Nowak and Cirpka, 2004). In closed form, the equation for the full posterior covariance matrix is

$$\mathbf{V} = \mathbf{G_{ss}} - \mathbf{G_{sy}} \mathbf{G_{yy}^{-1}} \mathbf{G_{sy}^{T}} \tag{3.43}$$

where $\mathbf{G_{sy}} = \mathbf{G_{ss}} \mathbf{H}^{T}$ and $\mathbf{G_{yy}} = \mathbf{H} \mathbf{G_{ss}} \mathbf{H}^{T} + \mathbf{R}$. In the case where compression of $\mathbf{Q}_{ss}$ is not used, the full matrix $\mathbf{V}$ is calculated and reported. Where compression of $\mathbf{Q}_{ss}$ is used, however, the diagonal of $\mathbf{V}$ is returned as a vector of variances on parameters. This information is reported in a separate file but is also used to calculated posterior 95 percent confidence intervals. The full matrix, when reported, can be used to calculate conditional realizations (Kitanidis, 1995, 1996).

References Cited

Box, G., and Cox, D. R., 1964, An analysis of transformations: Journal of the Royal Statistical Society, series B (Methodolodical), v. 26, no. 2, p. 211–252.

D'Oria, M., 2010, Characterization of aquifer hydraulic parameters—From Theis to hydraulic tomography: Università degli Studi di Parma, Ph.D. dissertation, 153 p.

Fienen, M., Kitanidis, P., Watson, D., and Jardine, P., 2004, An application of Bayesian inverse methods to vertical deconvolution of hydraulic conductivity in a heterogeneous aquifer at Oak Ridge National Laboratory: Mathematical Geology, v. 36, no. 1, p. 101–126, doi:10.1023/B:MATG.0000016232.71993.bd.

Fienen, M.N., Clemo, T.M., and Kitanidis, P.K., 2008, An interactive Bayesian geostatistical inverse protocol for hydraulic tomography: Water Resources Research, v. 44, W00B01, doi:10.1029/2007WR006730.

Gaganis, P., and Smith, L., 2001, A Bayesian approach to the quantification of the effect of model error on the predictions of groundwater models: Water Resources Research, v. 37, no. 9, p. 2309–2322, doi:10.1029/2000WR000001.

Gallagher, M.R., and Doherty, J., 2007, Parameter interdependence and uncertainty induced by lumping in a hydrologic model: Water Resources Research, v. 43, no. 5, W05421, doi:10.1029/2006wr005347.

Gray, R., 2005, Toeplitz and circulant matrices—A review: Delft, The Netherlands, Now Publishers, 90 p.

Hoeksema, R.J., and Kitanidis, P.K., 1984, An application of the geostatistical approach to the inverse problem in two-dimensional groundwater modeling: Water Resources Research, v. 20, no. 7, p. 1003–1020, doi:10.1029/WR020i007p01003.

Kitanidis, P.K., 1995, Quasi-linear geostatistical theory for inversing: Water Resources Research, v. 31, no. 10, p. 2411–2419, doi:10.1029/95WR01945.

Kitanidis, P.K., 1996, Analytical expressions of conditional mean, covariance, and sample functions in geostatistics: Stochastic Hydrology and Hydraulics, v. 10, no. 4, p. 279–294, doi:10.1007/bf01581870.

Kitanidis, P.K., and Vomvoris, E.G., 1983, A geostatistical approach to the inverse problem in groundwater modeling (steady state) and one-dimensional simulations: Water Resources Research, v. 19, no. 3, p. 677–690, doi:10.1029/WR019i003p00677.

Li, W., and Cirpka, O.A., 2006, Efficient geostatistical inverse methods for structured and unstructured grids: Water Resources Research, v. 42, no. 6, W06402, doi:10.1029/2005WR004668.

Li, W., Englert, A., Cirpka, O.A., Vanderborght, J., and Vereecken, H., 2007, Two-dimensional characterization of hydraulic heterogeneity by multiple pumping tests: Water Resources Research, v. 43, no. 4, W04433, doi:10.1029/2006WR005333.

Nowak, W., and Cirpka, O.A., 2004, A modified Levenberg-Marquardt algorithm for quasi-linear geostatistical inversing: Advances in Water Resources, v. 27, no. 7, p. 737–750, doi:10.1016/j.advwatres.2004.03.004.

Nowak, W., Tenkleve, S., and Cirpka, O.A., 2003, Efficient computation of linearized cross-covariance and auto-covariance matrices of interdependent quantities: Mathematical Geology, v. 35, no. 1, p. 53–66.

Press, W.H., Teukolsky, S.A., Vetterling, W.T., and Flannery, B.O., 1992, Numerical recipes in C—The art of scientific computing (2d ed.): Cambridge, UK; New York; Cambridge University Press, 994 p.

Rubin, Y., 2003, Applied stochastic hydrogeology: Oxford, UK; New York; Oxford University Press, 391 p.

Appendix 4—Parallel Implementation of Jacobian Calculations

Introduction

A substantial computational burden in quasi-linear parameter estimation is the repeated calculation of the Jacobian matrix (**H**). Calculation of this matrix is "embarassingly" or "pleasingly" parallel—the number of runs required is the same as the number of parameters, but the individual model runs do not need to interact in any way. Given this property of Jacobian matrix calculation, a cluster of individual computers can be used to make the computations in parallel. Parallel implementation can be performed using a problem-specific implementation coded into the software (for example, Schreüder, 2009), compiling the software against parallelization libraries (for example, Muffels and others, 2012), or using a parallel management system outside the software (for example, Condor Team, 2012).

As an initial step toward full support of parallel Jacobian calculations, this appendix documents an implementation using the Condor package (Condor Team, 2012) and Python scripts. This implementation currently calculates the Jacobian matrix using forward finite-differences with the increment set by the user in the derinc variable in the parameter_groups table (appendix 1).

Requirements

Several requirements must be met to use this parallel implementation for Jacobian calculations.

1. Condor and the related program DAGman must be installed and active on both the master node where bgaPEST is to run and at least one worker node. Details for installation are available from Condor Team (2012).

2. Python, including the module numpy, must be installed on all compute nodes that will be used by Condor, including the master node where bgaPEST is running.

3. In the folder where bgaPEST is running, the following files must be present

   - Condor_ATC.py
   - parallel_condor_Jacobian.py
   - jacobian_pre.py
   - jacobian_pre.bat (or jacobian_pre.sh on non-Windows operating systems)
   - jacobian_post.py
   - jacobian_post.bat (or jacobian_post.sh on non-Windows operating systems)
   - condor_jacobian.sub.orig
   - unzip.exe
   - zip_results.py

4. A folder called "data" must be present in the same folder where bgaPEST is running on the master node. The "data" folder must contain all forward-model related files, the .bgp file, and all .tpl and .ins files used by bgaPEST. Additionally, the following files must be present

   - tempchek.exe (included in example files or available from http://www.pesthomepage.org)
   - inschek.exe (included in example files or available from http://www.pesthomepage.org)
   - condor_single_run.py
   - parallel_condor_Jacobian.py

5. Set the variable deriv_mode=4 in the algorithmic_cv Keywords input block (appendix 1). This instructs bgaPEST to write necessary data transfer files and to execute the code necessary to run the Condor-based parallel Jacobian calculation.

File modifications

The file `condor_jacobian.sub.orig` must be customized to contain instructions specific to the Condor network on which bgaPEST is running for the `requirements` and the `request_memory` variables. Details of these variables are described by Condor Team (2012). All other variables should remain unchanged from the files provided with this documentation.

Description of the Method

This parallel implementation of bgaPEST using Condor is a scripted approach using a combination of existing PEST utilities and custom Python code. The general approach consists of the following steps:

1. Files are written by bgaPEST to communicate to the parallel codes information specific to the current run including derivative increments (specified by the user in the `derinc` variable).

2. At the time of each Jacobian calculation, the current parameter values are written to a temporary file

3. The current parameter values are combined with all other necessary model files in the "data" subfolder, which is compressed into a zipfile.

4. The Condor submit file is updated to specify the correct number of model runs required to calculate the Jacobian matrix.

5. A single-node directed acyclic graph (DAG) is initiated using the Condor utility `DAGman`. The need for the DAG is to monitor the set of Condor jobs corresponding to the specific model runs. `DAGman` monitors the progress of the DAG and returns control to bgaPEST ocne the entire Jacobian is calculated.

6. On each worker node, the following tasks are performed:

   - Model input files are written by using the `tempchek.exe` utility
   - The model is run once using the model command line information provided by bgaPEST
   - The model output files are read by using the `inschek.exe` utility. The version of `inschek.exe` provided with bgaPEST has been modified to carry maximum numerical precision.

7. After the model is run for each perturbation and a base case, the derivatives are calculated using forward differences and conveyed back to bgaPEST by using a text file.

References Cited

Condor Team, 2012, Condor Version 7.6.6 Manual: Madison, Wisconsin, University of Wisconsin—Madison.

Muffels, C., Schreüder, W., Doherty, J., Karanovic, M., Tonkin, M., Hunt, R., and Welter, D., 2012, Approaches in highly parameterized inversion—GENIE, a general model-independent TCP/IP run manager, U.S. Geological Survey Techniques and Methods, book 7, chap. C6, 26 p.

Schreüder, W., 2009, Running BeoPEST, *in* Proceedings, PEST Conference 2009, Potomac, Md., November 1–3, 2009: Bethesda, Md., S.S. Papadopulos and Associates, p. 228–240.

Appendix 5—Single-Layer Example Application

The first example application presented is a single-layer groundwater model. The forward model is a steady-state MODFLOW-2005 model with 21 rows and 21 columns and with constant row and column spacing of 1 meter (m). The hydraulic conductivity field is heterogeneous, varying from 6.068×10^{-5} to 0.048 meters per day. The true hydraulic conductivity field is depicted in figure 5.1. Observations of head at the locations shown in figure 5.4 were used for parameter estimation. To generate observations representing what would be field measurements in a non-synthetic case, the model was run forward using the true synthetic hydraulic conductivity values and the resulting head values were perturbed with normally distributed noise with mean of zero and standard deviation of 0.01 m.

The boundary conditions are constant head, highest at the northwest corner and linearly decreasing to the southeast, as depicted in figure 5.2. There is a well at row 13, column 6, extracting water at a constant rate of 0.231 liters per minute. No recharge is simulated in this case.

Two options exist to calculate sensitivity (Jacobian) matrices: (1) an experimental adjoint-state version of MODFLOW or (2) finite difference calculations using the Python linkage and PEST, as capable in the released version of bgaPEST. Two scenarios were tested as well: (1) a case in which the epistemic error term $\left(\sigma_R^2\right)$ is estimated and (2) one where σ_R^2 is fixed at a value approximately 2 orders of magnitude higher than the artificial noise used to corrupt the synthetic "true" values previously generated by the model—1.0 m. This level of epistemic uncertainty is intended to be unrealistically high, but it encompasses both measurement and modeling error and was thus used to demonstrate a case where overfitting would be avoided at all costs.

Figures 5.3 and 5.4 show the estimated parameter field and the corresponding squared residuals, respectively, for the case in which σ_R^2 is fixed at 1.0 m. In this case, the relatively high value ascribed to epistemic uncertainty certainly prevents overfitting. The solution, in fact, is smooth, and the linear variogram slope parameter θ was estimated to be 3.66×10^{-1}. Appropriately, the most structure expressed in the parameter field and the correspondingly lowest residuals are found near the pumping well where stress is the greatest (and therefore the amount of information is the greatest) is found. This phenomenon is discussed in further detail by Fienen and others (2008).

The smoothness of the parameter field is as expected from the bgaPEST algorithm and consistent with the maximum entropy property of the algorithm. In other words, the true hydraulic conductivity field is rough but the algorithm estimates parameters that smooth over the rough areas. The "smearing" of higher values across a larger area than in reality is also consistent with both maximum entropy and information content—an anomalous region of high hydraulic conductivity emanates from the area of the well in response to the focus of stress (and therefore information) being near the pumping well.

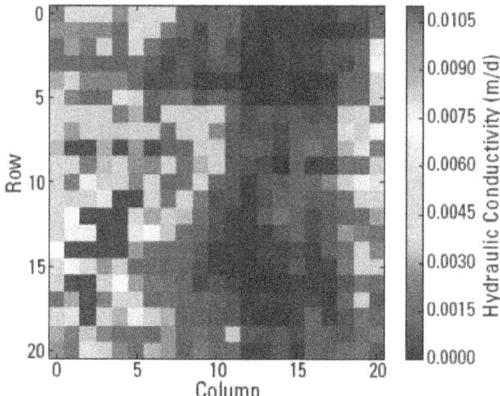

Figure 5.1.   True hydraulic conductivity field for the single-layer example application.

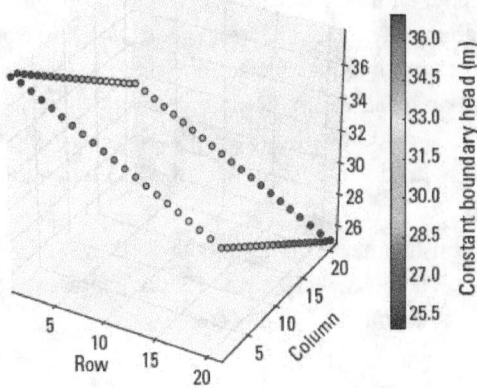

Figure 5.2. Boundary condition (constant head) for the single layer example application.

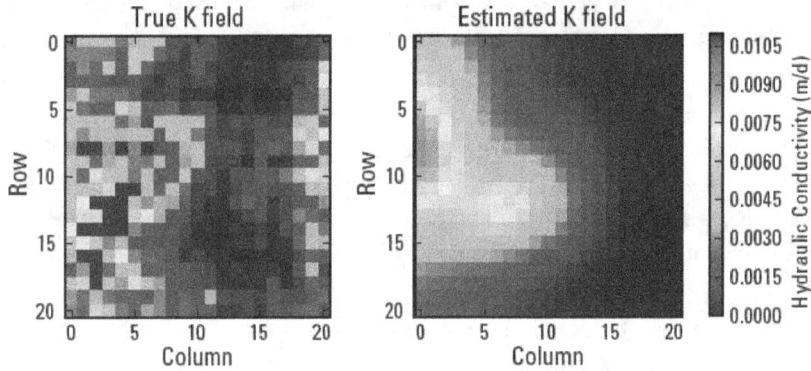

Figure 5.3. Parameters (hydraulic conductivity field) estimated for the single-layer example application using bgaPEST with σ_R^2 fixed at 1.0 m.

Figures 5.5 and 5.6 show the estimated parameter field and the corresponding squared residuals, respectively, for the case in which σ_R^2 is estimated by the bgaPEST algorithm. The estimated value for epistemic uncertainty was 1.007×10^{-1} and the variogram slope (θ) was 2.836×10^{-1}. These results are similar to the case in which σ_R^2 was held constant. The parameter field shows a similar shape and smoothness level, although figure 5.5 shows a bit more structure (roughness). Correspondingly, the residuals are generally lower. A key point here, however, is that the pattern of the residuals is similar and, again, reflects the general information content of the stress induced on the system.

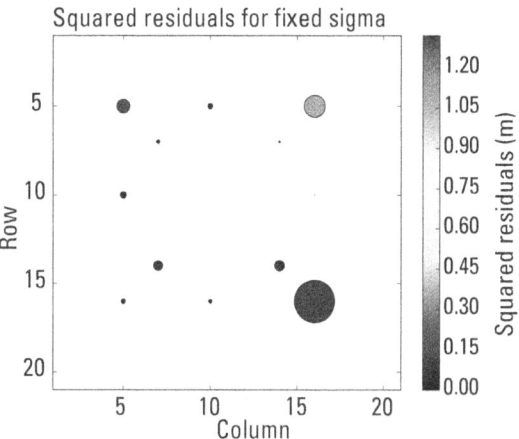

Figure 5.4. Squared residuals, plotted at their locations in the model, for the 1 layer example application using bgaPEST with σ_R^2 fixed at 1.0 m.

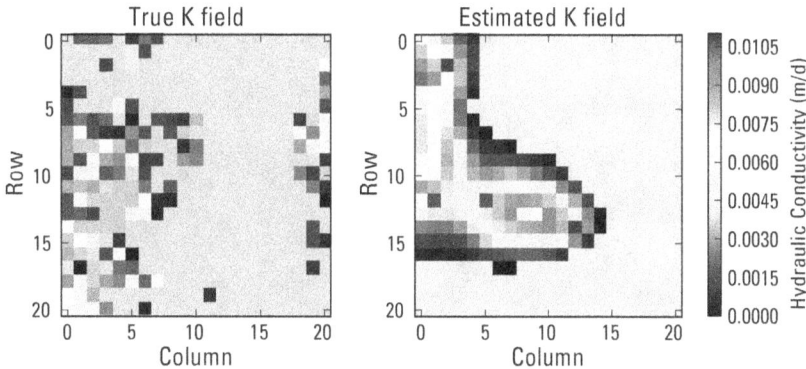

Figure 5.5. Parameters (hydraulic conductivity field) estimated for the single-layer example application using bgaPEST with σ_R^2 estimated by the bgaPEST algorithm.

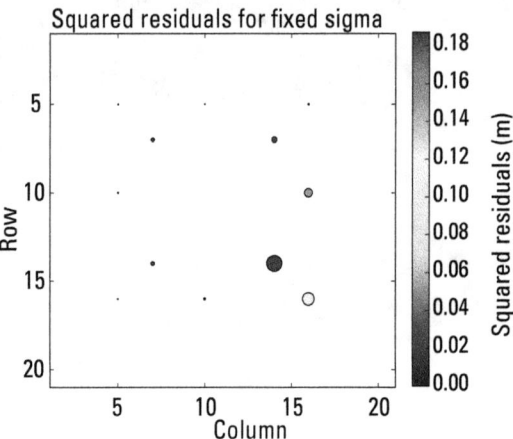

Figure 5.6.   Squared residuals, plotted at their locations in the model, for the single-layer example application using bgaPEST with σ_R^2 estimated.

Reference Cited

Fienen, M.N., Clemo, T.M., and Kitanidis, P.K., 2008, An interactive Bayesian geostatistical inverse protocol for hydraulic tomography: Water Resources Research, v. 44, W00B01, doi:10.1029/2007WR006730.

Appendix 6—Three-Layer Example Application

A three-layer groundwater model is presented to explore the use of multiple beta associations, anisotropy, and a larger number of parameters. In this case, the model is 40 rows by 35 columns across 3 layers. The row spacing is 2.0 meters (m) while the column spacing is 1.5 m. The layers, from shallowest to deepest, are 1.8 m, 1.4 m, and 1.8 m in thickness, respectively. The disparate row, column, and layer spacing was used to test the Toeplitz compression option. The model has constant head boundaries on all sides (set at the same elevation—60 m) and a single well at row 18, column 17, extracting at 0.01 liters per minute from each layer. This low flow rate is not meant to represent typical field conditions but rather highlights what can be learned with even a very small stress on the system.

The true parameter field, shown in figure 6.1, varies from 0.01 to 0.075 meter per day.

Five cases are illustrated here, as summarized in table 6.1. In all cases, each layer is treated as a separate beta association. In each of these layers, the initial value for the prior structural parameter (the linear variogram slope, θ) is 1.0×10^{-5}.

Cases 1 and 2 illustrate how the level of fit (and, therefore, the degree of roughness of the solution) can be influenced by adjusting the epistemic uncertainty term $\left(\sigma_R^2\right)$; so in these cases, σ_R^2 is set at a fixed value. In case 3, σ_R^2 is set low $\left(1.0 \times 10^{-5}\right)$ and the restricted maximum likelihood algorithm is given the freedom to estimate it. This setup illustrates the best achievable fit that one might achieve given the specific observation set provided without regard for overfitting. In cases 5 and 6, estimates are made with specification of

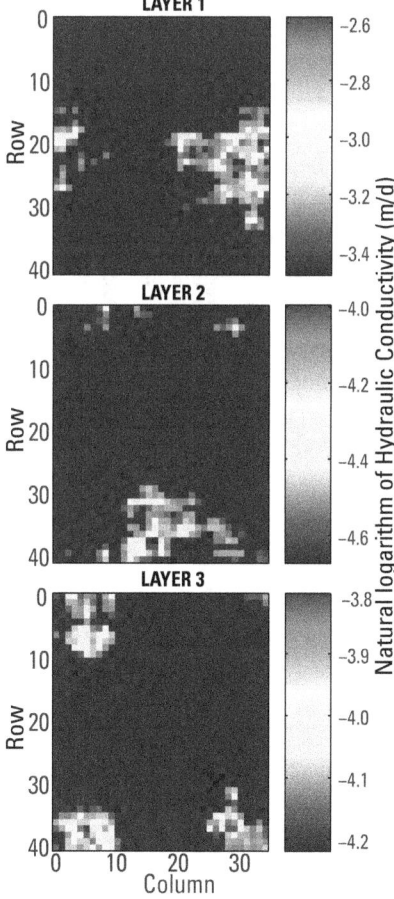

Figure 6.1. True synthetic hydraulic conductivity field for each layer in the three-layer example application. Values are shown in natural logarithm space to make the differences more visible.

Table 6.1. Summary of the five cases investigated. The table shows which structural parameters were estimated and fixed, and also indicates anisotropy when used.

		Scenario	Case 1	Case 2	Case 3	Case 4	Case 5
Prior Parameters		Initial σ_R^2	1.00E-01	1.00E-02	1.00E-05	1.00E-04	1.00E-01
		Estimated σ_R^2	-	-	7.79E-08	1.18E-05	-
	Beta Association 1	Initial θ	1.00E-05	1.00E-05	1.00E-05	1.00E-05	1.00E-05
		Estimated θ	2.46E-03	1.55E-02	1.25E-02	5.54E-03	3.61E-03
	Beta Association 2	Initial θ	1.00E-05	1.00E-05	1.00E-05	1.00E-05	1.00E-05
		Estimated θ	6.16E-03	2.47E-02	1.34E-02	3.19E-03	7.97E-03
	Beta Association 3	Initial θ	1.00E-05	1.00E-05	1.00E-05	1.00E-05	1.00E-05
		Estimated θ	2.46E-03	1.55E-02	1.21E-02	2.51E-03	7.73E-05
Anisotropy Parameters	**Beta Association 1**	horiz_angle	-	-	-	0.0	0.0
		horiz_ratio	-	-	-	100.0	100.0
		verical_ratio	-	-	-	1.0	1.0
	Beta Association 2	horiz_angle	-	-	-	0.0	0.0
		horiz_ratio	-	-	-	100.0	100.0
		verical_ratio	-	-	-	1.0	1.0
	Beta Association 3	horiz_angle	-	-	-	0.0	0.0
		horiz_ratio	-	-	-	100.0	100.0
		verical_ratio	-	-	-	1.0	1.0

anisotropy in the prior covariance. Inspection of the true parameter field in figure 6.1 suggests a possible correlation along the horizontal axis, indicative of a channel feature. In cases 5 and 6, therefore, an arbitrarily chosen ratio of 100 is applied with a rotation angle of zero. In case 4, like in case 3, σ_R^2 is estimated to achieve the best possible fit, whereas in Case 5, σ_R^2 is held constant at 1.0×10^{-1}.

Figures 6.2 and 6.3 show the estimated hydraulic conductivity field and squared differences between measured and observed head values, respectively, for case 1. In this case, meant to be conservative with respect to overfitting, the squared differences are smaller in magnitude than the specified value of σ_R^2 $\left(1.0 \times 10^{-1}\right)$ and very little roughness in the solution is required to achieve the level of fit desired.

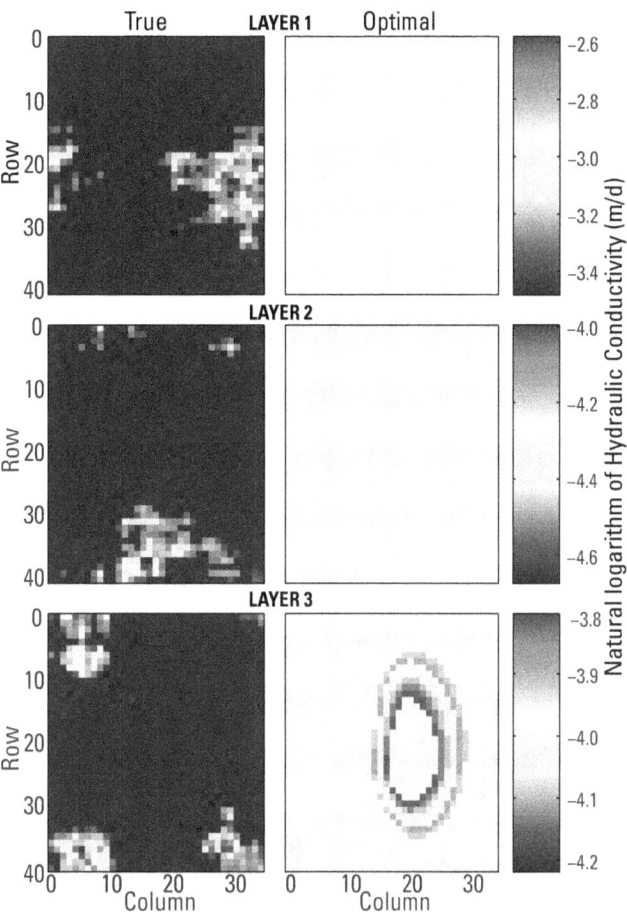

Figure 6.2. Case 1: Hydraulic conductivity fields estimated by using bgaPEST compared to the true, synthetic hydraulic conductivity field. σ_R^2 is held constant at 1.00×10^{-1}.

Figures 6.4 and 6.5 show the estimated hydraulic conductivity field and squared differences between measured and observed head values, respectively, for case 2. In this case, the specified value of σ_R^2 $\left(1.0 \times 10^{-2}\right)$ is lower than in case 1 and, accordingly, the squared head differences are lower, and more structure (roughness) is observed in the parameters, as expected. Note that, in this case, even with very low residuals, the parameter fields estimated are a smoothed representation of the "truth."

Figures 6.6 and 6.7 show the estimated hydraulic conductivity field and squared differences between measured and observed head values, respectively, for case 3. In this case, the value of σ_R^2 is estimated by the restricted maximum likelihood value algorithm. The head values match perfectly to machine precision, and the roughness of the field is the greatest of cases 1 through 3, as expected. The major features of the "true" hydraulic conductivity field are reproduced by this solution although they are smoothed, somewhat, as expected. Importantly, although the highest hydraulic conductivity values in layer 2 are slightly offset to the west, no artifacts are introduced that would be considered spurious in this solution.

Figures 6.8 and 6.9 show the estimated hydraulic conductivity field and squared differences between measured and observed head values, respectively, for case 4. In this case, the value of σ_R^2 is set very low $\left(1.0 \times 10^{-4}\right)$ to attempt to achieve excellent fit while introducing anisotropy with the principal direction aligned with the horizontal axis. In layer 1, a somewhat spurious artifact is visible in the form of a high hydraulic conductivity zone near the middle of the field. The head targets almost match within machine

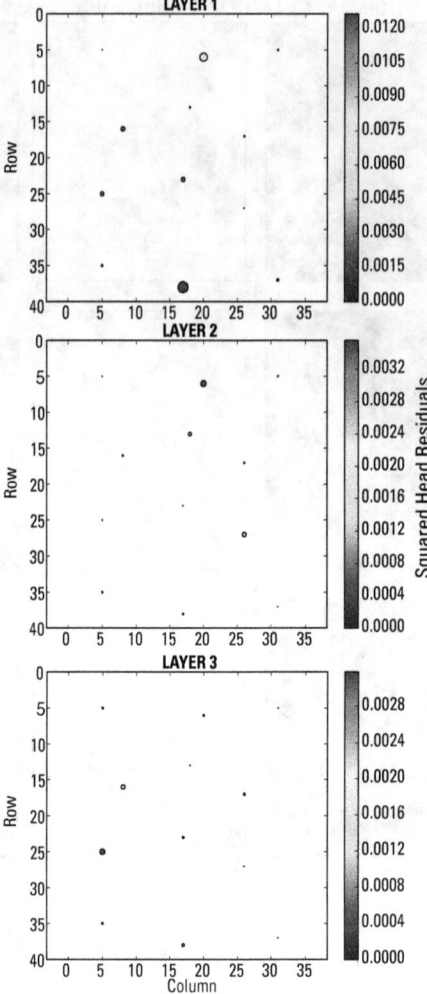

Figure 6.3.   Case 1: Squared differences between modeled and "true" head values. Symbol size qualitatively indicates magnitude, and color scale quantifies magnitude. Locations of the circles indicate observation location in the model domain in plan view. σ_R^2 is held constant at 1.00×10^{-1}.

precision, however, and all other features are reasonable. This highlights the fact that, within a single beta association, if anisotropy is used, *all* features estimated will roughly correspond to that framework so, in a Bayesian sense, the answer is *conditional* on the prior assumption that the anisotropy is an appropriate general characteristic shape of the parameter field. Such assumptions must be made cautiously.

Figures 6.10 and 6.11 show the estimated hydraulic conductivity field and squared differences between measured and observed head values, respectively, for case 5. In this case, the value of σ_R^2 is set at a the same value as case 1 $\left(1.0 \times 10^{-1}\right)$ to compare a solution with and without anisotropy assumed. Because anisotropy is a reasonable characteristic of the "true" field in this case, better fits are achieved (nearly an order of magnitude lower residuals) and the general pattern of the parameter field is better in case 5 with anisotropy than in case 1 without anisotropy. This highlights the power that anisotropy can bring to a parameter estimation problem when it is appropriate even when σ_R^2 is set conservatively to avoid overfitting. As discussed above, however, this anisotropy will, in a sense, force the solution to conform to such a shape, so its use should be approached with caution.

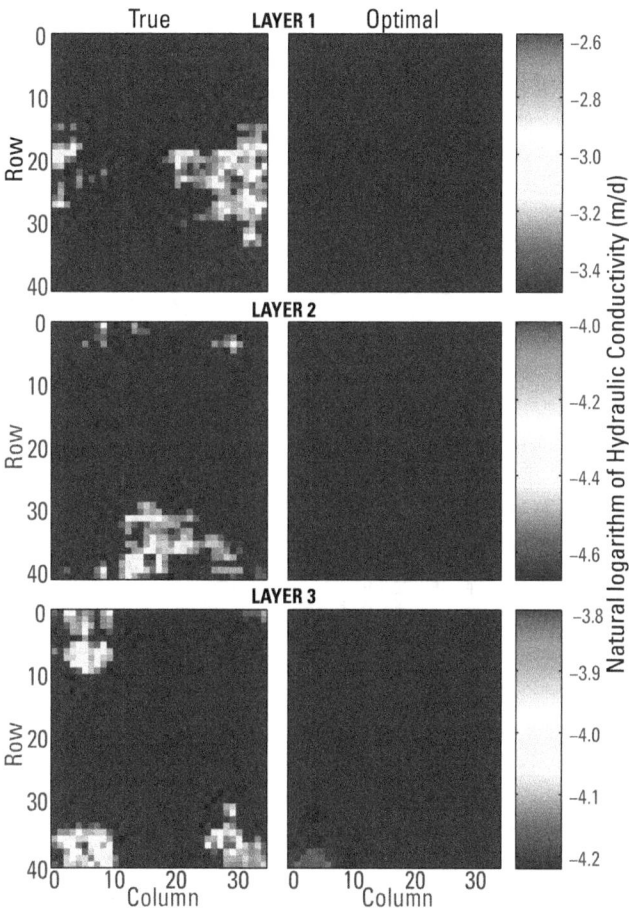

Figure 6.4. Case 2: Hydraulic conductivity fields estimated by using bgaPEST compared to the true, synthetic hydraulic conductivity field. σ_R^2 is held constant at 1.00×10^{-2}.

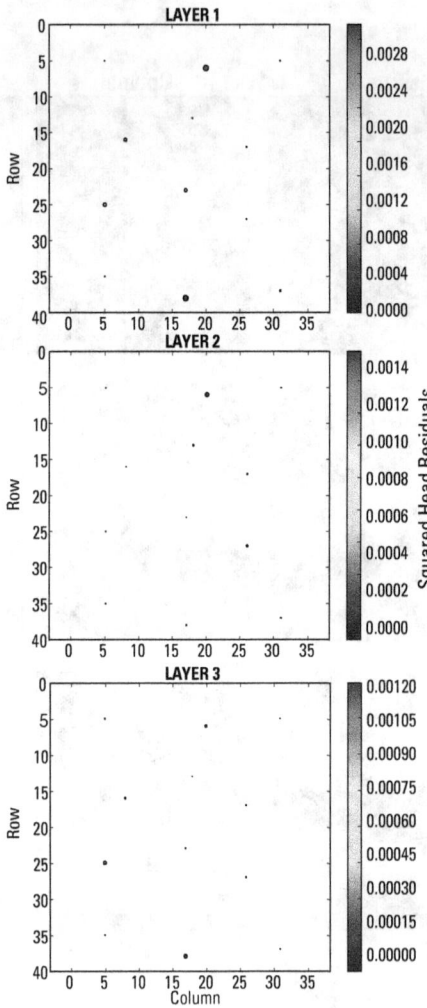

Figure 6.5.   Case 2: Squared differences between modeled and "true" head values. Symbol size qualitatively indicates magnitude, and color scale quantifies magnitude. Locations of the circles indicate observation location in the model domain in plan view. σ_R^2 is held constant at 1.00×10^{-2}.

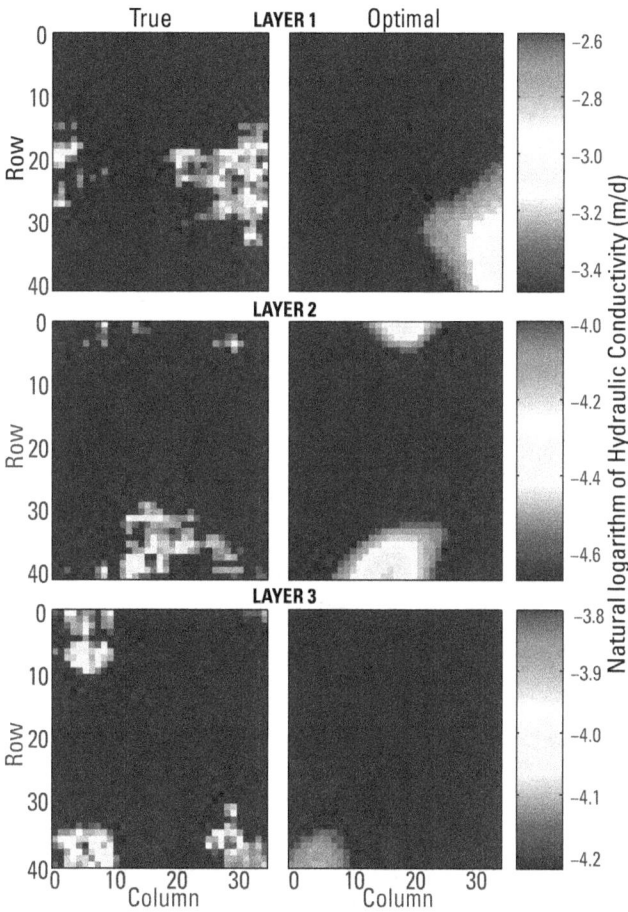

Figure 6.6.   Case 3: Hydraulic conductivity fields estimated by using bgaPEST compared to the true, synthetic hydraulic conductivity field. σ_R^2 is initially 1.00×10^{-5} and estimated by bgaPEST at an optimal value of 7.79×10^{-8}.

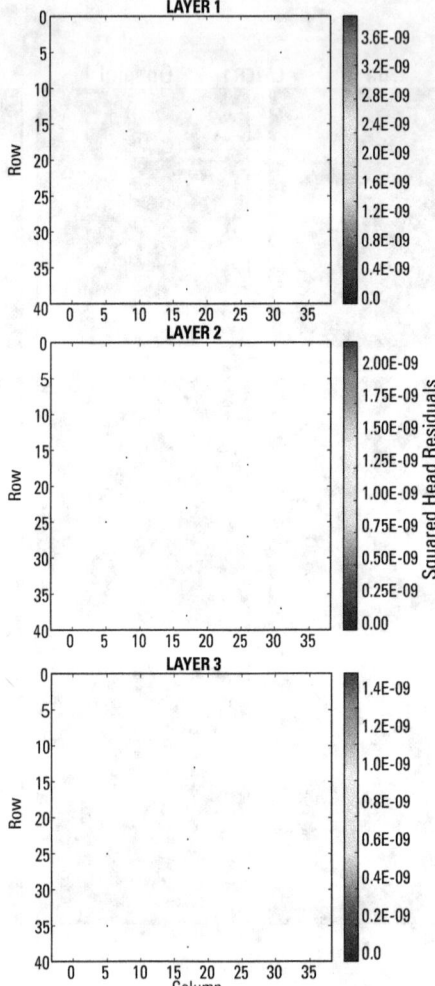

Figure 6.7.   Case 3: Squared differences between modeled and "true" head values. Symbol size qualitatively indicates magnitude, and color scale quantifies magnitude. Locations of the circles indicate observation location in the model domain in plan view. σ_R^2 is initially 1.00×10^{-5} and estimated by bgaPEST at an optimal value of 7.79×10^{-8}.

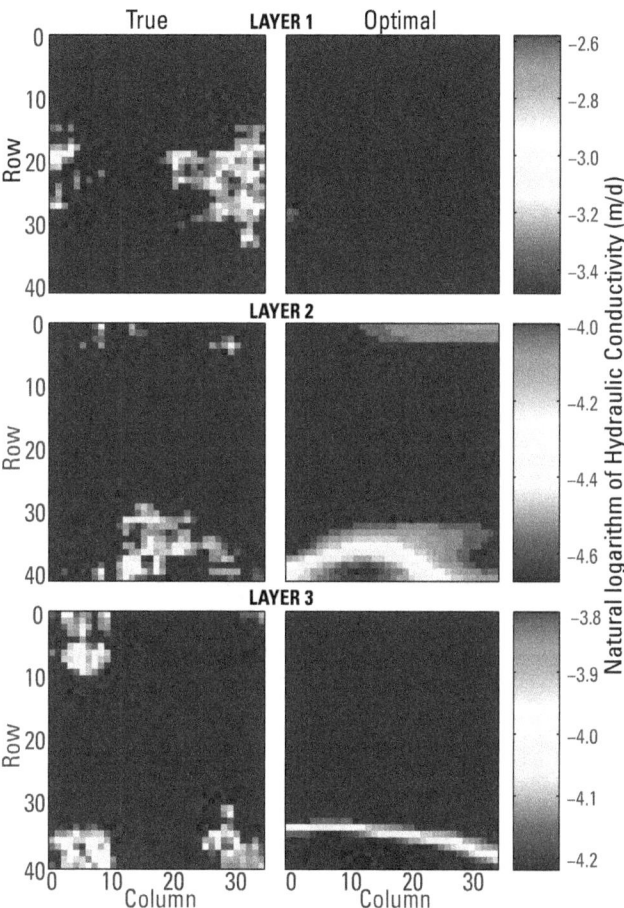

Figure 6.8.   Case 4: Hydraulic conductivity fields estimated by using bgaPEST compared to the true, synthetic hydraulic conductivity field. σ_R^2 is initially 1.00×10^{-4} and estimated by bgaPEST at an optimal value of 1.18×10^{-5}. Parameter anisotropy also invoked as described in Table 6.1.

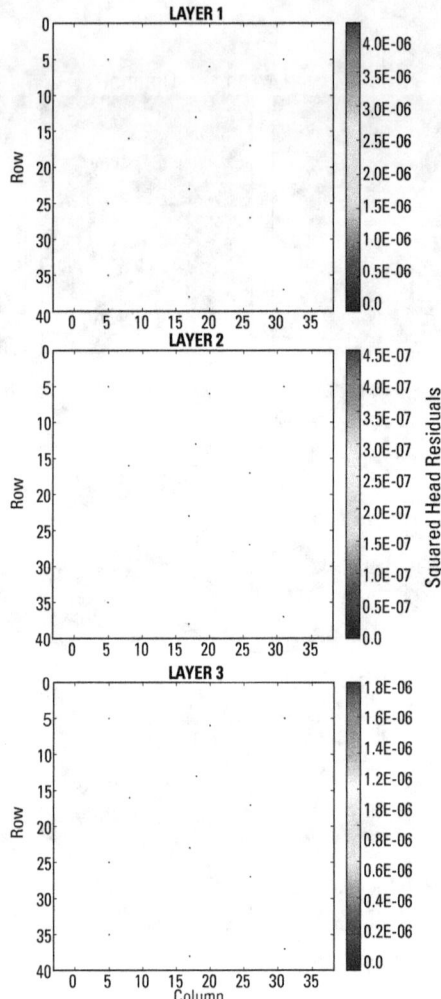

Figure 6.9.   Case 4: Squared differences between modeled and "true" head values. Symbol size qualitatively indicates magnitude, and color scale quantifies magnitude. Locations of the circles indicate observation location in the model domain in plan view. σ_R^2 is initially 1.00×10^{-4} and estimated by bgaPEST at an optimal value of 1.18×10^{-5}. Parameter anisotropy also invoked as described in Table 6.1.

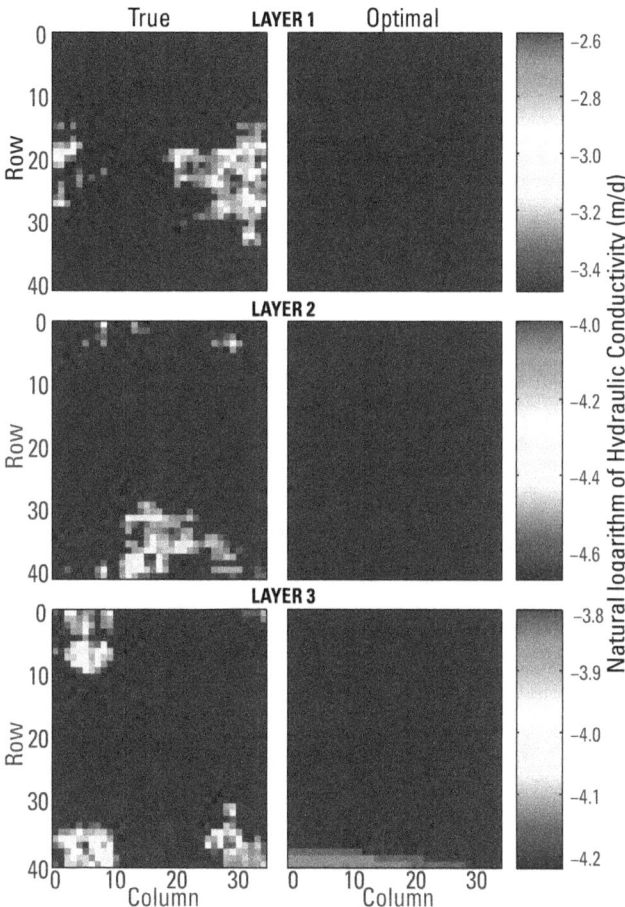

Figure 6.10.   Case 5: Hydraulic conductivity fields estimated by using bgaPEST compared to the true, synthetic hydraulic conductivity field. σ_R^2 is held constant at 1.00×10^{-1}. Parameter anisotropy also invoked as described in Table 6.1.

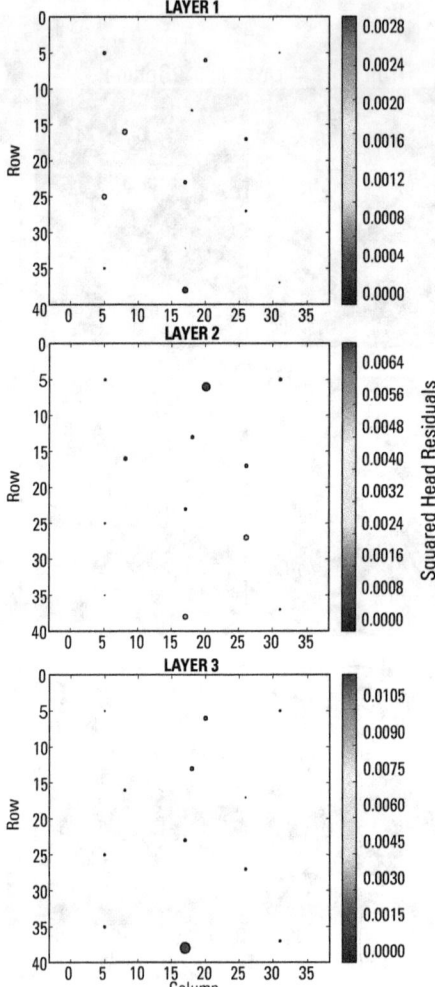

Figure 6.11.   Case 5: Squared differences between modeled and "true" head values. Symbol size qualitatively indicates magnitude, and color scale quantifies magnitude. Locations of the circles indicate observation location in the model domain in plan view. σ_R^2 is held constant at 1.00×10^{-1}. Parameter anisotropy also invoked as described in Table 6.1.

Appendix 7—Reverse Flood Routing Example Application

An example application of reverse flood routing in open channels is presented in this section (D'Oria and Tanda, 2012). Reverse flood routing is useful to obtain hydrographs at upstream ungaged stations by means of information available at downstream gaged sites. The considered channel was prismatic and 20 kilometers long; the cross sections (spaced by 100 meters (m)) were trapezoidal in shape with bottom width of 10 m and side slope of 2/1. A longitudinal channel slope of 0.001 and a unitless Manning coefficient of 0.033 were adopted. The Manning coefficient is expressed as unitless when all other quantities are in SI units (as is the case here). A correction factor would be required if English units were used for the other quantities. The upstream and downstream boundary conditions were a streamflow time series and the uniform flow condition, respectively. The initial condition was set consistent with the steady state of a constant flow rate equal to the first value of the upstream hydrograph. The BASEChain module of BASEMENT (Faeh and others, 2011) that solves the De Saint Venant equations for unsteady one dimensional flow was adopted as forward model. A flood wave with time to peak of 2.5 hours, peak flow of 164 cubic meters per second (m^3/s), and base flow of 25 m^3/s was considered to obtain the corresponding downstream outflow subsequently corrupted with multiplicative random errors and used in the inverse procedure. The simulation time was equal to 15 hours; the input and output hydrograph time discretization was constant and equal to 5 minutes resulting in 181 values. The initial flow condition and the downstream streamflow time series (181 observations) were then used to estimate the inflow hydrograph (181 parameters). The initial parameter values were set to the mean value of the observations. The sensitivity (Jacobian) matrix was evaluated by means of a finite-difference calculations using the Python linkage and PEST, as capable in the released version of bgaPEST. The epistemic error term σ_R^2 and the linear variogram slope parameter θ were estimated by the restricted maximum likelihood value algorithm.

In figure 7.1 the actual input hydrograph, the actual downstream hydrograph, assessed by applying the forward model, and the error-corrupted one used for the inversion are reported along with the reproduced inflow and the corresponding outflow. Table 7.1 summarizes the estimated structural parameters. In the first case the downstream hydrograph was corrupted with a 1 percent multiplicative random error (figure 7.1a), in the second case a 10 percent multiplicative random error was used (figure 7.1b). In both the inversions, there is a close match between the estimated input hydrograph and the actual one; the peak flow and time are properly reproduced. The estimated epistemic error variance increases in the second case taking account of the higher erroneous observations (table 7.1).

References Cited

D'Oria, M., and Tanda, M.G., 2012, Reverse flow routing in open channels—A Bayesian geostatistical approach: Journal of Hydrology, doi: 10.106/j.hydrol.2012

Faeh, R., Mueller, R., Rousselot, P., Vetsch, D., Volz, C., Vonwiller, L.R.V., and Farshi, D., 2011, System manuals of BASEMENT, version 2.1: Zurich, Switzerland, ETH Zurich Laboratory of Hydraulics, Glaciology and Hydrology (VAW).

Table 7.1.   Reverse routing: estimated structural parameters .

Random errors	1%	10%
θ [m^{6}s^{-3}]	2.0×10^{-2}	1.0×10^{-2}
σ_R^2 [m^{6}s^{-2}]	1.7×10^{-1}	9.3

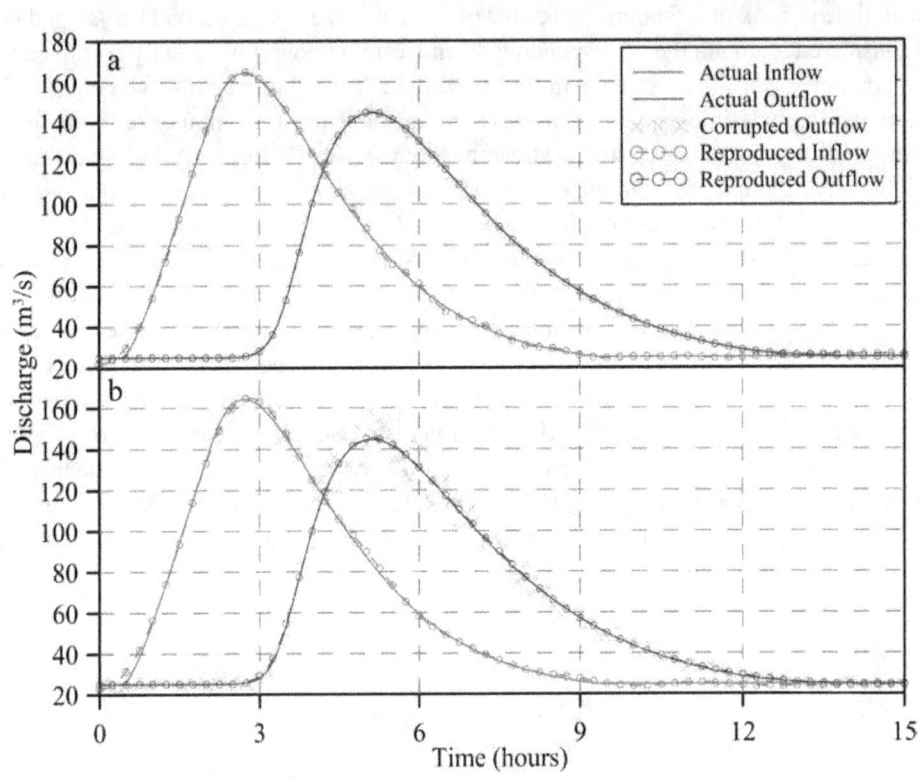

Figure 7.1.   Reverse routing: inflow and outflow hydrographs for the prismatic channel. (a) the observations were corrupted with a 1 percent random error; (b) a 10 percent random error was used.

Glossary

Mathematical Symbols and Variables

m — Number of parameters

n — Number of observations

p — Number of beta associations

$\mathbf{s}$ — Vector of parameters

$\mathbf{s}_0$ — Starting values of parameters

$\mathbf{y}$ — Vector of observations

$\mathbf{y}_k'$ — Observations corrected for the kth linearization: $\mathbf{y}_k' = \mathbf{y} - \mathbf{h}\tilde{\mathbf{s}}_k + \tilde{\mathbf{H}}_k\tilde{\mathbf{s}}_k$

$\mathbf{h}(\mathbf{s})$ — Modeled results of $\mathbf{s}$

$\mathbf{H}$ — Jacobian matrix of parameter sensitivities

$\mathbf{W}$ — Diagonal weight matrix

σ_R^2 — Epistemic uncertainty

σ^2 — Variogram variance or slope parameter

$\mathbf{R}$ — Epistemic uncertainty covariance

$p(\mathbf{s})$ — Prior probability of parameter values

$L(\mathbf{y}|\mathbf{s})$ — Likelihood function

$p(\mathbf{s}|\mathbf{y})$ — Posterior probability of parameter values

$\mathbf{X}$ — A selection matrix that can include drift of the prior mean

β — Prior mean values for each beta association

β^* — Initial prior mean values for each beta association used in the prior pdf

β_0 — Starting prior mean values for each beta association

θ — Vector of the structural parameters

$\hat{\theta}$ — Best estimate of the vector of the structural parameters

θ^* — Initial value of structural parameters (used in the prior pdf of θ)

θ_0 — Starting value of the structural parameters

$p(\theta)$ — Prior probability of structural parameter values

$L\left(\mathbf{y}_k'|\theta\right)$ — Likelihood function in structural parameter optimization

$p\left(\theta|\mathbf{y}_k'\right)$ — Posterior probability of sturctural parameters given the current data

$\mathbf{G}_{ss}$ — Prior covariance of $(\mathbf{s}-\mathbf{X}\beta^*)$

$\mathbf{G}_{sy}$ — Cross covariance between observations and parameters

$\mathbf{G}_{yy}$ — Auto-covariance of the observations when covariance of the prior mean is included

$\mathbf{Q}_{ss}$ — Prior covariance of the parameters

$\mathbf{Q}_{yy}$ — Auto-covariance of the observations

$\mathbf{Q}_{\beta\beta}$ — Prior covariance of the prior mean values

$\mathbf{Q}_{\theta\theta}$ — Prior covariance of the structural parameter values

$\tilde{\mathbf{s}}$ — Current best estimate of $\mathbf{s}$ during quasi-linear runs

$\tilde{\mathbf{H}}$ — Current sensitivity matrix, a function of $\tilde{\mathbf{s}}$ during quasi-linear runs

ξ — Interpolation weights for the innovation in solving for $\hat{\mathbf{s}}$

$\hat{\mathbf{s}}$ — Best estimate of $\mathbf{s}$

$\hat{\beta}$ — Best estimate of β

Φ_T — Total objective function

Φ_M — Measurement component of objective function

Φ_R — Regularization component of the objective function

Φ_S — Structural parameters objective function

$\mathbf{s}_{\mathbf{opt}}$ — Optimal value of parameters in the linesearch

ρ — Linesearch parameter

$E[\cdot]$ — Expected value

$\mathbf{d}$ — Separation distance in variogram calculation

ℓ — Integral scale in variogram calculation

$\mathbf{v}$ — Vector of epistemic uncertainty terms

$\mathbf{W}$ — Observation weight matrix

$\mathbf{T}$ — An arbitrary block Toeplitz matrix

ω — Observation weight

$\mathbf{p}$ — Untransformed parameter values in the context of power transformation

α — Exponent for power transformation

$\mathbf{V}$ — Posterior covariance matrix of the parameters

www.ingramcontent.com/pod-product-compliance
Lightning Source LLC
Chambersburg PA
CBHW081550170526
45166CB00009B/2646